DAIRY FARMING

DAIRY FARMING:

WHAT COWS TO BUY, HOW TO HOUSE, FEED, AND CARE FOR THEM AND MAKE DAIRYING PAY

D.S. BURCH

Published by

FacsiMile Publisher
12 Pragati Market
Ashok Vihar, Ph-2
Delhi - 110052, India
Phone: 011-47034999
Email: www.facsimilepublishers.com
info@facsimilepublishers.com

Distribution Network
gyanbooks.com
India, USA, Canada, UK, Australia

ISBN: 978-93-6896-714-9 (PB)
First Published, 1912

2nd Impression 2025

Printed at: Gyan Press, Delhi.

DAIRY FARMING
Author: D.S. BURCH

I.—A PRACTICAL COMMERCIAL DAIRY BARN with two stave silos in the foreground. Note the large double doors in front and the ventilators in roof of the barn.

Dairy Farming

BY

D. S. Burch

State Dairy Commissioner of Kansas

WHAT COWS TO BUY, HOW TO HOUSE, FEED, AND CARE FOR THEM AND MAKE DAIRYING PAY

Philadelphia
The Penn Publishing Company
1912

Contents

CHAP.		PAGE
	INTRODUCTION, DAIRYING, THE INDUSTRY OF IMMEDIATE CASH RETURNS	7
I.	WHO MAY SUCCEED IN DAIRYING?	12
II.	HOW TO SELECT A DAIRY FARM .	15
III.	WHICH BREEDS ARE THE BEST AND WHY	20
IV.	HOW TO CHOOSE A DAIRY COW .	31
V.	HOUSING AN EQUIPMENT . .	42
VI.	SOLVING THE HELP PROBLEM .	58
VII.	POINTS ON MILKING . . .	65
VIII.	PRACTICABILITY OF MILKING MACHINES	68
IX.	FEEDING AND MANAGING THE HERD	78
X.	TESTING MILK FOR RICHNESS AND PURITY	89
XI.	SANITATION AND CARE OF MILK .	98
XII.	CERTIFIED MILK AND ITS PRODUCTION	107
XIII.	THE SCIENCE OF BOTTLED MILK .	111
XIV.	MILK INSPECTION AND ITS PURPOSE	116
XV.	DAIRY BUTTER—HOW TO MAKE THE BEST	119

XVI.	Is Farm Cheese Making Practical?	133
XVII.	The Truth About the Tuberculin Test	138
XVIII.	Dairy Organizations	149
XIX.	How and Why the Dairy Markets Fluctuate	157
XX.	Why Oleomargarine is Inferior to Butter	164
XXI.	The Creamery Promoter and His Tactics	169
XXII.	Dairy Farm Fertility	175
XXIII.	Dairy Accounts	182
XXIV.	Dairying for Health and Wealth	188
XXV.	Dairygrams, Proverbs of Unlimited Application	190
XXVI.	Some Common Dairy Terms	192
	Index	205

Dairy Farming

INTRODUCTION

DAIRYING, THE INDUSTRY OF IMMEDIATE CASH RETURNS

A cow's breath has been said to be the sweetest of all known things, and while such an abstract quantity cannot be compared with more concrete substances for relative sweetness, we must grant that in many ways the dairy cow is a remarkable and wonderful creature. Not only is this true, so far as we are able to measure her value in dollars and cents, but also her unique position as the foster mother of the human race is a position for which no other animal has yet been able to successfully compete. On the stock exchanges and in the great markets, we so constantly hear wheat and grains discussed that the uninitiated in agricultural matters are likely to form the opinion that grain farming is the chief source of agricultural wealth in the United States and the world.

It is not our purpose to emphasize one branch of agriculture at the expense of any other, but rather to advise beginners and those unacquainted with the various types of agriculture in order that they may wisely select the most remunerative branch. It has been said that the cow, the hog and the hen constitute the foundation of our national agricultural wealth. The dairy cow is rightly placed at the head of the trio; she gives approximately six hundred crops of milk a year, a product for which in its various forms there is an unlimited outlet.

The average man, woman and child in the United States consumes over seven-tenths of a pint of milk per day or its equivalent in the form of cream. Add to this the butter, cheese, condensed milk and sundry dairy products and you have a conception of the market open to the dairyman and of the possibilities for engaging in and developing special branches of the dairy industry.

Milk dealers of New York City are obliged to go as far as two hundred miles into the country to purchase sufficient milk to supply their trade, and cream is shipped as far as seven hundred miles into that metropolis. What is true of New York is, in a smaller way, true of all large cities.

In fact the situation is becoming one of transportation rather than of production, since dairymen living within a short distance of cities have long ago lost hope of being able to compete with the demand.

The Markets Are Consistently Good

In a similar way, creameries and cheese factories are unable to supply the demand for the manufactured products, butter and cheese.

Therefore thousands of pounds of butter and cheese must be placed in storage every year in the height of the season of best production in order that the public need not go butter or cheese hungry during the fall and winter months, when the factories are totally unable to meet the demand, owing to the small production of milk on the average farm during the winter months.

The consistently high prices of milk, butter and kindred dairy products testify to the continuous and never glutted market for milk and its products. The markets are furthermore near at hand and of a cash or monthly payment nature. Instead, as in the case of grains, of one crop a year dependent on the climate and the weather, the dairy cow yields her milk twice a day for three hundred or more days out of the year.

Throughout the length and breadth of the civilized world, a quart of fresh milk is practically legal tender and with almost as constant a value as gold. From any railroad station in the country, ship a ten-gallon can of rich cream to almost any one of our six thousand creameries, and a check for five dollars or more will be forthcoming in a few days, accompanied by a request for further shipments.

The Hand Separator Plays an Important Rôle

The general introduction of the small centrifugal cream separator for farm use has made possible a great development of dairying, for instead of shipping the bulky milk either by wagon or rail, the farmer is now able to reduce its volume about seven times and ship the cream, the most valuable part, to whatever market he may choose and retain the skim-milk on the farm for feeding purposes. Again, the rapid growth of the ice-cream industry offers a wide and favorable market for the dairyman who is in a position to produce pure sweet cream.

While the problems of production of farm products are now fairly well solved as a result of the labors of the state and federal scientists, profitable marketing of farm products has not yet been thoroughly worked out in a practical way—with the exception of the dairy industry. In this case, the market is practically constant and usually satisfactory. This advantage makes dairying the main prop of the renter of small means, the city man about to move to a farm and the farmer who requires a steady cash income throughout the year.

CHAPTER I

WHO MAY SUCCEED IN DAIRYING?

BEFORE investing capital or labor in any line of business, to which dairying is no exception, the prospective investor of course considers first the probable chances for success and the dividends which he may reasonably expect to secure on his investment. The question therefore "Who May Succeed In Dairying?" is vital.

Success is here considered in a financial way rather than in its more delicate ethical sense. Those who are best able to succeed in dairying should possess a general knowledge of all farm operations. In addition to this they must be familiar with the feeding and management of a dairy herd, even though help is hired to perform the actual work. They must have sufficient business ability to make the necessary purchases to the best advantage and to select the most profitable markets. They must be receptive of up-to-date methods, the demands of the trade, and the importance of a uniformly high quality of product.

They must have a reasonable amount of capital, land and equipment. They must understand and appreciate the importance of sanitation and be able to discriminate between the essential and the non-essential in operations bearing on sanitation. They should have a deep human interest in their work and, aside from a financial aspect, should take personal interest in every animal in the herd. The successful dairyman soon develops an affection for his well-groomed and soft-eyed cows.

Business Ability is Essential

While the wise selection of the farm, of stock and of hired help are important determining factors in success, the man who is a lover of animals has much in his favor. Any good business man or a student of the world may succeed in dairying, but a knowledge of the principles of animal breeding, and an understanding of the "why" of the various operations as well as the "how" are of inestimable benefit in reducing the number of errors in case one is a beginner.

On every dairy farm there are also a great many seemingly non-essential details which should be mastered before one actually engages in the business. Among such details may be mentioned the construction and repair of fences, best types of stalls, use of fly repellents in summer and care and handling of ugly bulls. While these subjects will be subsequently discussed as fully as possible, any one unfamiliar with dairy farm details and about to engage in dairying, can wisely afford to hire out for a few months even at very low wages.

Importance of Both Scientific and Practical Knowledge

In gaining experience in this way, he should, however, select a farm on which to work resembling his own or his prospective farm as closely as possible, both as to size and the kind of dairying conducted. A short term in one of the numerous agricultural colleges or dairy schools, now found in every state, will also be of value in a scientific as well as a practical way.

The foregoing remarks and qualifications apply chiefly to those of limited dairy experience. As the highest success in modern dairying involves both a practical and scientific knowledge, it is obvious that those with a wide practical experience should reinforce their knowledge from scientific sources, and those with a technical knowledge study the more practical side.

CHAPTER II

HOW TO SELECT A DAIRY FARM

On the selection of the dairy farm depends not only the success of the investment and the size of the financial returns but also, in the majority of cases, the personal satisfaction derived from living amid congenial surroundings. Following are a few essential points, gleaned from a wide field of experience, which require little comment, so self-evident is the dairy wisdom they contain.

1. Do not buy a farm because it is advertised as a bargain. Its cheapness is determined by a lack of demand and the lack of demand is generally based on one or more important objections which either make the farm profitless or uncongenial for human habitation.

2. Do not assume more than two-thirds indebtedness if you must depend on the farm for a living. In case you are unable, after securing the necessary stock and equipment, to pay at least one-third down, either defer your purchase or rent. Simple mathematical computations show the difficulty of meeting interest and paying off the principal where less than one-third cash is paid.

3. *Do not purchase a farm for dairy purposes without thoroughly investigating the water supply, both as to quantity and quality.* All live stock require an abundant supply of pure water. It is desirable that water be available in several different parts of the farm.

4. *Do not purchase a farm which is incapable of growing grass and such forage crops as clover or corn.* Pasture and good hay are almost indispensable to successful dairying under conditions prevalent in the United States.

5. *Be cautious in purchasing a mortgaged farm, or one which has been abandoned for a number of years.* A farm which has been mortgage-ridden for a considerable period of time has generally been cropped so heavily and avariciously that its productiveness is a very uncertain factor. In a similar way an abandoned farm is invariably overrun with obnoxious weeds which can be subdued only at great expense.

6. *Consider carefully the nature of the roads, traction facilities and the distance from permanent markets.* If milk is to be sold for city consumption it must be delivered daily and good roads

are, of course, essential. If milk or cream is to be sold to a creamery or cheese factory, the farm should be located within easy reach of several of these. In the present stage of dairy development an individual creamery or cheese factory is an un-

II.—A TYPICAL DAIRY FARM in Wisconsin showing milk house at extreme right, masonry silo at its left, dairy barn in center and house at extreme left.

certain market, as it may shut down permanently or at least close for the winter without giving notice. Where several creameries or cheese factories are within reach, the risk is reduced to a minimum. Creameries and cheese factories, however, are excellent markets in the respect that they will readily take all the product they can secure if the quality is satisfactory, since the larger the quantity of business they can do, the more economically they can operate.

7. *In determining the size of the dairy farm, allow three acres of the total land for each cow.* The actual amount of pasture may be about one acre for each cow, the remaining two acres being devoted to crops for winter feeding, reserve pasture, barn-yards and buildings. In Denmark, Holland, and certain other parts of Europe, where dairying is conducted on a very intensive basis, the cows are kept in the stable the greater part of the time. This practice enables more land to be devoted to forage crops and more cows may be kept. Imitations of this system have not been financially successful in the United States, owing to the much greater cost of hired help in this country. Pasturing eliminates, to a great extent, the hauling of feed to the barns, the hauling away of manure and the care incident to habitual stall feeding.

8. *Consider the opportunities you will have to secure hired help.* Dairy farming, especially when conducted on a large scale, involves the problems of securing and keeping competent and reliable help. Therefore, consider carefully the possibil-

ities of securing help locally, as bringing it in from the outside or advertising for it is not only expensive but is often unsatisfactory.

9. *Look into the social life of the community.* One's neighbors are more important in the country than in the city, and the selection of social life in the country is virtually a selection of the country itself. Learn the prevailing nationality of your neighbors and their ideals and consider the advantages or probable nature of your social life from the standpoint of your hired help as well as of yourself.

10. *Lastly, look into the legality of the title and secure expert advice as to the terms of purchase.* The legal details of transferring real estate and the basis for estimating land values, taxes and insurance vary in different states and communities. These should, of course, be investigated before, not after, the land has been purchased.

The foregoing rules and requirements may seem to the reader, at first glance, extremely rigid. Not so; they exclude only the worst farms, so many of which are a drug on the real estate market. Of good farms there are plenty, waiting only for brainy and energetic management.

CHAPTER III

WHICH BREEDS ARE THE BEST AND WHY

EVERY breed of cattle has been developed and perfected with a certain goal in view. That is, breeders have, by careful selection and mating, made cattle adapted for specialized purposes.

The following are the principal breeds of cattle and the purposes for which they are used:

Milk	*Milk and Beef*	*Beef*
Jersey	Shorthorn	Hereford
Guernsey	Brown Swiss	Aberdeen Angus
Holstein	Polled Durhams	Galloway
Ayrshire	Red Polls	Devons

The foregoing classification applies to the average type of cow of each breed. Individuals and special "strains" of any breed frequently show characteristics which place them in a class to which the breed as a whole does not belong.

The choice of breed for dairy purposes is generally made from the first column and wisely so, for these breeds have, for the past half century, attracted favorable attention as heavy milk producers. The center group includes breeds commonly known as general purpose or dual-purpose animals. Such breeds are adapted to the pioneer farmer or the man who engages in general farming and desires cows which require but a small amount of attention and still produce enough milk for the home table.

The dual-purpose animal has been criticized by many so-called farm specialists as being an attempt at a jack-of-all-trades in cow circles. Still, the valuable services which the breeds of the second group have given to the man engaged in general farming testify strongly to the merits and usefulness of these breeds in their proper place. In fact, more than seventy-five per cent. of the butter produced in creameries in the central West is derived from milk of dual-purpose cows, chiefly Shorthorns. The third group consists of breeds found chiefly on the great beef cattle ranches and stock farms, and may be dismissed from consideration by the dairy farmer.

The breeds in the first group from which the choice is generally made by experienced dairymen have the following origin and characteristics.

Jerseys

The **Jersey** cow originated on the Island of Jersey, in the English Channel, about thirteen miles from the French coast, and forty miles from the southern end of Great Britain. The Jerseys were recognized as a breed about the year 1750,

III.—A Promising Two-Year-Old Jersey.

but not till a century later did importations to the United States become active. Since then the breed has grown rapidly in popularity, and the name of Jersey cow is known wherever dairy cows are kept.

The chief characteristics of the Jerseys are the richness of their milk, their small size and docile disposition. Their milk is often too rich in fat for their own calves, but for a select market milk trade, Jersey milk is very desirable. The average weight of Jersey cows is about eight hundred and fifty pounds, and their small size results in economy of stable space. The bulls average about sixteen hundred pounds in weight. Pure-bred Jerseys vary in color, though shades of tan, fawn and yellow are the most prevalent. The head is small, eyes large and placid. The Jersey cow is light and graceful in movement, in fact, well-bred animals are distinctly aristocratic in their actions and appearance. Success with Jersey cattle depends largely upon proper housing and gentleness, as they are not as vigorous as some of the other breeds, and their nervous system is highly developed and sensitive. In the Jersey bulls, this characteristic of the breed is exhibited by fractiousness, especially when improperly handled, and old Jersey bulls frequently become ugly without apparent cause.

Jerseys, as a breed, are spare of flesh, a desirable dairy quality, but for this reason their carcasses are little valued for beef.

Guernseys

The Guernsey breed of cattle, which originated in the Island of Guernsey, also one of the Channel Islands, bears a strong resemblance to the Jerseys. In fact, both of these breeds are frequently referred to as Alderney cattle. Guernseys were recognized as a breed about 1850, and

IV.—Guernsey, showing typical markings.

importations to the United States began about 1875. The breed has found favor rapidly wherever introduced and properly handled.

Guernsey cattle are slightly larger and heavier

than Jerseys, somewhat coarser in appearance, though still very attractive. Their color is much the same as the Jerseys, though more often in patches, especially on the lower part of the body and legs. The cows give large quantities of milk, its richness, however, being somewhat less than that of Jerseys. Guernseys are gentle, quiet, easily managed, and these qualities are observable, to a less degree, even in the bulls. Both the Guernsey and Jersey breeds are noted for the rich yellow color of the milk and cream and butter made therefrom. The fat globules in their milk are relatively larger and separate more readily than those of other breeds of cattle.

Holsteins

The Holstein breed, though known in the United States under various names, is readily distinguished from other breeds by the characteristic black and white markings. In fact, they are frequently referred to in untechnical language as Blacks and Whites, although the more exact names are Holstein Friesians, Dutch Cattle or Holland Cattle.

As the last name implies, the cattle originated in Holland and from the beginning of the Christian era the breed has been constantly developed and improved. Owing to lowland pastures and the rich luxuriant grasses which have constituted the feed of Holstein cattle for centuries, they have developed large frames with a great capacity for milk production.

V.—A Well-Formed Holstein.

The Holsteins are perhaps the most picturesque of all dairy cattle. The Dutch make pets of their cattle, and as a result they have a quiet and docile disposition and are easily handled.

They are the largest of the dairy breeds and are

noted for remarkable records of milk production. The milk is less rich, both in total solids and the percentage of fat, than any of the principal dairy breeds.

The fat does not rise readily from the milk, owing to the small size of the fat globules. This characteristic of the milk makes it appear less rich than it actually is, but nevertheless, the consuming public is somewhat prejudiced against Holstein milk since they believe it less nourishing than a milk which produces more cream on standing or is of a yellower color. The chief influence which has lately begun to have a countereffect upon the public sentiment referred to has been the efforts of Holstein breeders along advertising lines to show the public that Holstein milk is very rich in tissue- and muscle-building properties. This claim is true, yet, were Holstein milk richer and of a higher yellow color, the breed would doubtless be a great deal more popular than it now is among producers of milk for city consumption.

As a matter of fact, the Holstein cow gives enough more milk than the other breeds to create a total butterfat production of very creditable amount. The world's record for the amount of butterfat produced in a year by a single animal has been held for several years by Holstein cows.

Ayrshires

The Ayrshire breed originated in and is named after the county of Ayrshire in southern Scot-

VI.—The Ayrshire, alert and active.

land. The country is rolling and the climate rugged. As a result Ayrshire cattle possess more stamina, as a breed, than any of the others mentioned. The breed has developed and has been raised into prominence since about the year 1825.

The most striking characteristic of the Ayrshire

cow is her ability to thrive on coarse feed and rough and scantily covered pasture-land. In size and weight the Ayrshire cow slightly exceeds the Jersey. The former is fine-boned, active, alert, but inclined toward a nervous temperament and is less docile than any of the dairy breeds. In color the typical Ayrshire is reddish brown splashed with white. The head is relatively small and the expression bright, almost to wildness. The horns have an upward growth with the tips inclined backward.

Compared with the other breeds, the Ayrshire cow has a small udder and teats. Breeders have been able to remedy these objections and a large number of Ayrshire herds now have udders and teats of normal size, though not large. Ayrshires are good milkers, but the milk is neither of exceptional richness nor is the cream of the rich yellow color usually demanded by the trade.

The relative merits of the four chief dairy breeds may be summed up as follows:

CHARACTERISTICS OF MILK

	Richness	*Amount*	*Color*	*Keeping Qualities*
1—	Jersey	Holstein	Jersey	Ayrshire
2—	Guernsey	Guernsey	Guernsey	Holstein
3—	Ayrshire	Ayrshire	Ayrshire	Guernsey
4—	Holstein	Jersey	Holstein	Jersey

CHARACTERISTICS OF ANIMALS

	Gentleness of Cows	*Gentleness of Bulls*	*Constitutional Vigor*	*Relative size of Cows*	*Value for Beef*
1—	Jersey	Holstein	Ayrshire	Holstein	Ayrshire
2—	Guernsey	Ayrshire	Holstein	Guernsey	Holstein
3—	Holstein	Guernsey	Guernsey	Ayrshire	Guernsey
4—	Ayrshire	Jersey	Jersey	Jersey	Jersey

CHAPTER IV

HOW TO CHOOSE A DAIRY COW

In the selection of cows for the dairy herd it is generally customary to look chiefly for productiveness, since productiveness is considered equivalent to profit. To consider only the amount of milk which a cow may produce is not, however, dairy wisdom. The chief consideration in the selection of a dairy cow should be her health, since the health of any animal affects not only her own value but the value of the entire herd. One of the most serious factors in the spread of the terrible scourge, tuberculosis, has been the indiscriminate and general introduction of diseased animals into healthy herds.

Tuberculosis cannot be readily detected by a superficial examination of the animal in the early stages of the disease, yet the animal may at the time of the inspection be expelling the invisible disease germs. Tuberculosis is not the only menace to dairy herds, but its importance is so great that every bovine introduced into the herd or brought on the premises for any purpose should either be isolated for a period of at least thirty days, during which the tuberculin test may be applied, or else should be accompanied by a certificate from a reputable veterinarian showing that the animal is in a healthy condition. The problem of controlling tuberculosis in the herd and the method of making the tuberculin test will be subsequently discussed.

The Principle of Cross-Breeding

The second consideration of importance in selecting cows for the dairy herd should be their blood origin. There is neither profit nor satisfaction in conducting a dairy composed of "scrub" animals or mixed breeds, however well their performances at the pail at any particular time seem to justify their place in the herd. It is noteworthy that the first cross between pure-bred animals is often very satisfactory as a milk producer and as a valuable asset to the herd. But, when the second cross is made, the amount of milk produced generally diminishes and the animal is of inferior quality in other respects. If cross-breeding is further continued, the dairyman who practices it will ultimately have a herd of cows whose value as breeding animals is very uncertain.

Contrast with this condition of affairs the experiences of wise dairymen, who have started their herds with a few pure-bred animals and have followed a systematic breeding campaign. Their first cows may perhaps not have been unusually large producers, yet, if the sire was from a family of profitable cows, the offspring will be at least equal and probably superior in value to the average between the sire and dam. This truth is based on the well-recognized principle that nature aspires to perfection.

Pure-bred Stock is Always in Demand

Further breeding will result in more pure-bred animals, and in view of the fact that a pure-bred herd requires no greater amount of expense and attention for care and maintenance than one of nondescript animals, the breeder will, in a few years, be well repaid financially for the line of breeding which he has followed. The problem of disposing of the male animals will be entirely solved, since pure-bred males are in good demand for breeding purposes, whereas ill-bred bulls of the dairy breeds are almost unsalable even to the butcher.

The female calves of pure-bred origin have in addition to their probable value as future milk producers the additional value of potential breeders. A study of the advertising columns of dairy papers will strongly impress upon the person about to engage in dairying the value of pure-bred stock. It matters not whether the preference has been given to the Holstein or Jersey, to the Ayrshire or Guernsey; the ultimate returns and the satisfaction of blooded animals will be equal. Self-made men may be in demand, but the poorly bred dairy cow, however phenomenal her performances may be, is not sought after, because breeders have learned that her milk-producing characteristics and capacity will probably disappear, on breeding, for several generations and the chances of securing valuable animals from such a cow are too slight to encourage the experiment.

Young Blood is Best for Foundation Stock

A third important factor in the selection of a dairy cow is her age. However attractive the cow may be in conformation, in style and from the point of view of past records, if she is old, she is not desirable as an addition to the herd. An old animal is very likely to be a poor breeder, and expensive to feed. When sent to the block her value will be very slight and in a great many ways she is undesirable. Young blood, particularly good young blood, is as desirable in the dairy herd as in all lines of business.

The factor of productiveness which one is so likely to consider first now arises. The true production of a dairy cow can be determined accurately only by means of the scale and the test. The scale tells us the amount of milk she produces and the test furnishes information as to the richness of the milk. For the market milk trade the amount of milk is perhaps the most important, but it must be remembered that the certain standards of richness must be complied with and the test, therefore, should by no means be overlooked. In producing milk for butter or cheese factories, both factors must always be considered, since such factories commonly pay in proportion to the number of pounds of butterfat (also called milkfat) delivered.

Forcing Cows to Make Records

Milk condenseries pay in accordance with the number of pounds of milk delivered, although the contract which they have with the producers generally specifies that the milk must be of certain richness, four per cent. being the usual standard. The records of production which give the true value of the dairy cow must extend over a period of at least one year. Some breeders, by giving their animals stimulating foods and milking them three times a day, are able to produce very exceptional weekly or monthly records.

They seldom, however, furnish the sequel to these records, for the reason that the period following is generally one of recuperation rather than of a creditable performance. Records extending over a short period of time afford an index to the capability of an animal, but are by no means as valuable as records of a year's duration or more.

Where records are unobtainable, one must rely upon the appearance of the cow in making his selection. Like a horse-trade, this manner of buying animals is generally in favor of the seller, but fortunately a dairy cow has certain indications of milk production which cannot be covered up or exaggerated. A good milking animal should show a tendency toward spareness and under no consideration should she carry a very large amount of fat. The dairyman wishes his feed to be converted into milk and not into beef.

Physical Points of a Good Milker

Looking at the animal from the head, one should seek and should find a broad forehead with large mild eyes set far apart. The neck should be thin and free of hanging fat. The chest should be well developed, yet not prominent. The body of the cow, commonly referred to as the barrel, should have a large capacity, as it is there that the feed is, through a wonderful chemical process, assimilated and transferred into blood and milk. The hip-bones should be prominent and far apart. A large capacity in the pelvic region indicates not only the probability of the cow being a large milk producer but also of being a good breeding animal.

Now looking at the cow from the side, she should appear straight on top, but with the lower line sloping toward the rear and suggesting a wedge. She should show no indications of being leggy, and in the terminology of good dairymen "there should be little sunlight under her." Both fore-legs and the hindquarters should be well muscled and sinewed but free from any more flesh than is necessary to comfortably carry the weight of the animal. From the rear the cow should show a deep, wide, and a well developed udder or bag. The teats should be placed far apart and should be of good size. The tail should be long and slender in preference to being short and stocky. The udder should be hung quite high and from the back of the flanks in preference to being attached to merely the under side of the animal.

Expert dairymen claim that if the skin of a dairy cow is soft, pliable, and oily she is likely to be a better milker than if it is hard, thick and unpliable. While opinions may be to the contrary, the writer is confident that quality and texture of the skin is at least an indication of excellence in general. The claim that yellow secretions of a waxy or oily nature, which may be found in the ears and near the roots of the hair, indicates richness and a yellow color in milk is unfounded.

Study the Animal's Disposition

The practical dairyman will of course milk a cow at least once before purchasing her. A hard milker or an animal with an undesirable disposition and a tendency to kick is an animal whose purchase should be considered at least twice, however strong may be her qualifications in other respects.

It must be remembered that a cow must be milked before she will yield a profit, and she must be milked not once or twice but twice a day, month after month and year after year.

Considerations of Minor Importance

Another consideration in the purchase of a dairy cow is her color, and her similarity to other animals in the herd. While a herd similar in color and markings is desirable from the standpoint of uniformity, which is an important factor in making sales, color should not be a matter on which to be too critical provided one is satisfied that her breeding has been along the desired lines.

Holstein cows, for example, are on the average about half white and half black, yet some of the most noted individuals depart from this average to such an extent that they are almost totally white or totally black. The same lack of uniformity of color applies to the other breeds as well, though not quite so strikingly perhaps. Since dairy cows of the same breed show wide differences in color or markings, this seemingly important but really trivial consideration should not affect the purchase of an otherwise meritorious animal.

The Range of Prices for Desirable Animals

In regard to the prices which one must expect

to pay, information can be given in a general way only, due to fluctuating factors of supply and demand, local conditions and current prices of feeding materials. Pure-bred cows with satisfactory records for milk production generally change hands for from eighty to two hundred dollars. If an animal has been registered, the price is somewhat larger. Grade cows bring from forty to one hundred dollars, and scrubs, or those of nondescript breed, seldom exceed forty dollars in value.

The average value placed upon milch cows by government statisticians has ranged from thirty-five to fifty dollars during the past ten years. It is obvious that the great majority of dairy cows in the United States are poorly bred animals owing to the unfortunate lack of information prevalent among farmers on questions of breeding.

For a number of years the price of milch cows has been steadily increasing, due largely to the better prices obtainable for dairy products. Following are the principal factors which increase and decrease the value of milch cattle.

Factors Which Elevate the Value of Dairy Animals

Certificate showing that the animal has successfully passed the tuberculin test.

Authentic records showing creditable yield of milk and butterfat.

An age of between two and eight years.

A record for being a successful breeding animal.

A satisfactory pedigree.

Good crops, general prosperity and abundant feed at low prices.

High prices for dairy products in immediate vicinity or in markets easily reached.

Factors Which Reduce the Value of Dairy Animals

The converse of all the above factors.

Extreme old age, exceeding eleven years.

Inability to secure hired help for milking.

The shutting down of near-by creamery or cheese factory or failure of any other important market near.

A serious epidemic in the vicinity or likelihood of the presence of tuberculosis in the herd.

Strict enforcement by the dairy authorities of legal requirements to which a considerable percentage of the animals in the vicinity are unable to conform.

CHAPTER V

HOUSING AND EQUIPMENT

THE dairy business is no exception to the rule that profits cannot be made without an investment; still, by careful planning, one is able to reduce his outlay for buildings and equipment to a very modest sum. The value of a dairy barn for housing milch cows depends not upon elaborate systems of automatically feeding and caring for the cows, not upon the number of sacks of cement used in its construction, nor upon the skill with which the architect planned its artistic appearance. The value of a dairy barn depends upon the efficiency of the ventilating system, upon the means which have been devised for reducing the amount of dust to a minimum, upon the number and size of the windows and upon well arranged and comfortable stalls for the cows.

THE BASEMENT BARN IS OBJECTIONABLE

Barn plans are not difficult to make, and it is far better to plan the barn for the particular kind of dairying to be conducted rather than to decide upon a particular plan and make the business conform to it. Types of barns vary in different climates, but for all purposes, those found in the central West are superior to those of any other great section. The basement cow stable so prevalent in the eastern states is perhaps the most objectionable, and few of them are now built.

VII.—WAITING TO BE MILKED.—A herd of pure-bred Guernseys in a well-drained barn-yard.

Basement stables with their low ceilings, damp floors and dark corners are ideal habitations for all the bacterial diseases to which cows are subject, and are extremely unpleasant to work in.

The stable should be slightly above the level of the surrounding ground, well drained and easily approachable from the pasture, without requiring the cows to wade through mud and filth just before milking time.

Concrete or Wood?

The relative merits of concrete and of wood construction will be a matter for the dairyman to consider, together with the price of each. Concrete construction is ideal for the foundation and the lower portion of the barn, owing to the ease with which it can be cleaned and also to the fact that it is rat-proof. Concrete construction is, at the present time, rather expensive, and no doubt always will be. Wood, on the other hand, is a perishable building material, but for the purpose of the average dairyman it serves and satisfies every reasonable need.

A wooden barn built on a solid concrete foundation is perhaps the most economical and satisfactory where a dairy farm must be put on a paying basis in a short time. There are so many smaller and important necessities, such as watering tanks, feeding floors, gutters and the like, that where economy must be considered, it is far better to have the latter of concrete than to make the barn of concrete and to use inferior material for the sundry equipment.

When a Cement Floor is Permissible

The stable floor and the stable ceiling are two very important considerations in the ideal dairy barn. Except in the extreme southern latitudes, cows should not be required to stand or lie on cement. Cement is not only slippery when wet but it is likely to chill the cow's sensitive milk secreting glands if the udder comes in contact with the floor, especially during the night, when the ground is cool. However much bedding is used, this objection cannot be overcome. A cement floor is desirable for cow stables provided it is covered with a removable wooden panel or false floor on which the cow stands or lies. The gutter, back of the platform on which the cow stands, should be constructed of concrete with a slope of about one inch in ten feet. Careful dairy farmers who recognize the value of the liquid manures for fertilizing purposes construct cement cisterns under the ends of the gutters to catch and hold this important by-product of the dairy herd.

Floor plans may be made so that the cows either face a center feeding alley or so that they face the walls and the space between them can be used for removing manure and carrying the milk away. Both types of construction have their advantages and one must decide whether he prefers greater facilities for handling the feed or for handling the milk and removing the manure easily. The writer personally prefers the cows to face a center feeding alley, for in this way the animals are able to see each other and when confined in the barn for a considerable portion of the day, as during the winter months in northern latitudes, the animals are generally more contented.

The Cow-Barn Ceiling

The ceiling of the dairy barn should be at least nine feet above the floor level and should be dust- and chaff-proof. Gentlemen farmers prefer not to store hay, grain and other feeds above the stables. This necessitates bringing the feed from other buildings and obviously greatly increases the amount of work required for feeding the animals. Where the ceiling above the stable is tight and impervious to dust, there can be no

VIII.—A MODERN DAIRY BARN.

practical objection to storing feed over the stable. Hay can readily and with little labor be passed down into the stable through chutes and grain can be economically handled by means of grain boxes which connect bins above the stable with the lower floor. The ceiling should be kept well whitewashed in order that dust, cobwebs, and dirt of all kinds may be readily detected, also for the additional light afforded by reflection from the white surface.

Sunlight as a Germicide

Considering the sides of the stables, the matter of windows and doors is the most important. Doors should be at least three and one-half feet in width, opening outward, and should be of liberal height, at least six feet, three inches. Many a valuable cow has been injured by crowding through a narrow door. There should be no step up or down for the cow to take when entering the stable, since a slight slip may strain the delicate organs in which the milk is formed and secreted. The windows should aggregate at least ten per cent. of the floor space. Abundant light is not only nature's best germicide but makes the long hours spent in the dairy barn more pleasant for both men and animals. The walls in front of the cows, in case they face toward the outside, should be painted or in some way colored a modest brown, drab, or other semi-dark hue. Experience has shown that nervousness or excitability in dairy animals is increased if they are obliged to face a bright-colored or a whitewashed wall.

VENTILATION

Ventilation of the dairy barn by other means than through windows and doors is desirable. Ventilating shafts a foot or two from the floor and extending to the roof where they open into a cupola are the simplest ventilators. Another excellent system is to have cold air inlets under the windows, with an exit for the heated and foul air near the ceiling. At least one square foot of ventilator space is necessary for every three animals kept, if reasonable purity of the air is expected.

Drinking water is sometimes provided in the stable, but in case a sheltered watering tank in the yard can be arranged, the latter method is preferable under average conditions.

Stalls and Stanchions

Stanchions are undoubtedly the best means of fastening the cows while in the stable and while being milked. These may be made of wood (and come within the province of any carpenter familiar with farm work) or they may be of metal, a large variety of which are on the market and advertised in prominent farm papers. A stanchion is superior to a rope and halter since the cow has a freer use of her head in a stanchion, is held more firmly should she prove fractious and there is no rope to interfere with her eating.

The chief requisites of manufactured stanchions should be to allow the cows to lie down comfortably, to be easily opened and closed and quickly cleaned. In addition to the stanchions for the cows in milk, there should be provided three or four comfortable box stalls for every twenty-five animals, in order that suitable quarters may at all times be furnished cows just before and during calving. The box stalls should be about eight feet square and should preferably be situated in the quietest place in the barn.

Suitable quarters for young calves must also be considered in the planning of the stable in case there is to be only one barn for the cows. Where several cow stables are to be built, a separate barn for the calves is desirable. A great deal of difficulty in managing calves at feeding time can be avoided by providing a row of stanchions for them. By so doing, the customary scuffle and the stealing of the smaller calves' feed by the larger ones will be avoided.

The Bull-Pen

As every dairyman with a herd of cows exceeding twenty should keep his own bull, a stout pen or box stall must be provided for "his majesty." The walls of the pen should be high enough so that the sire is not able to see the cows while they are in the barn.

While the cow stable is the center of operations on the dairy farm, other buildings are necessary for the proper care of milk and its products. The number of other buildings depends, of course, on the disposition made of the milk. Certified milk production requires, perhaps, the greatest additional outlay in the nature of buildings and equipments.

Separate Buildings for Special Operations

Where butter or cheese is made on the farm,

the number of buildings is somewhat less, and if the milk or cream is sold directly to creameries, cheese factories or city milk dealers, one building in which to care for milk and cream will suffice. The dairy laws of several of the most important states prohibit the keeping of a cream separator in the barn, unless the separator room is tightly boarded off from the stable. This precaution prevents the absorption of barn odors by the milk to a considerable degree. But every dairyman who wishes to put out a first-class product must remove the milk from the barn very soon after it is drawn, and the other operations which follow should be conducted in an entirely separate building.

Such a building or buildings should be at least seventy-five feet distant from the barn and should be connected with it by either cement, gravel or cinder walks. If the milkers carry mud and filth into the building where the milk is handled, the value of having the separate building is greatly reduced. A great many well-equipped dairy farms provide a trolley-carrier for removing the milk from the barn. Such a means of doing away with the necessity of the milkers or others who work in the dairy barn entering the building is an important step toward safeguarding the flavor of the milk. The equipment of the buildings for butter making, cheese making or for certified milk will be subsequently discussed under chapters dealing with these lines of dairying.

Silos and Silage

An almost indispensable building on the dairy farm is the silo. A silo is nothing more than an immense tank for preserving the succulent forage crops, together with all their nutritious juices, for feed during the winter months. In the extreme south, where green fodder is obtainable all the year around, silos may be dispensed with, but they are nevertheless of great value in storing up, in all climates, surplus corn fodder, cow-peas, green alfalfa hay and other well-relished feeds from years of plenty to years of scarcity. In the northern climates, where green feed is unobtainable during the winter months, also in time of serious drought, the value of the silo is becoming appreciated more and more every year. Silos are made in various sizes, shapes and styles. They are made of masonry, concrete, tile, cement blocks, wooden staves, ordinary wood framework, in fact

IX.—One Hundred Ton Concrete Silo erected at a cost of $293, of which $159 was paid for labor. The actual cost of the material, including $5 for

of every common building material. Of all shapes, the round silo is preferable, since the silage settles better and there is a greater capacity in proportion to the materials used in its construction than in any other shape.

Dimensions and Capacities

The dimensions of a silo depend chiefly on the rapidity with which silage will be fed. For a herd of twenty cows the silo should be twenty-four feet high and eighteen feet in diameter. Such a silo will hold slightly over one hundred tons of corn fodder. The silage is put into the silo in the fall by means of an ensilage cutter provided with a blower or other carrier to elevate the chopped green material to doors at the top of the silo from which it falls to the bottom. The silo should be filled in as short a time as possible in order that there may be no chance for decay or serious fermentation aside from the natural chemical changes which occur. The top three inches of the silage will always spoil unless fed immediately after the silo is filled, and feeding once begun must be continued to avoid loss by fermentation. In practice, the top is allowed to spoil, and when systematic feeding is begun it is discarded. During the cold months fermentation is very slight; but in summer, silage must be fed rapidly and constantly in order to prevent its spoiling on top.

Points in Silo Construction

Concrete and masonry silos are the most expensive of all types. Their expense is partly compensated for, however, by the very slight losses due to fermentation and decay as compared with silos of wood construction. Regardless of the material of which the silo is made, the depth should be at least one-third more than the diameter in order that the silage may settle solidly and force out the air. The smallest size that a silo should be built is fifteen feet in diameter and twenty feet deep; the largest size is about twenty-five feet in diameter and thirty-five feet deep. The latter size will hold approximately three hundred and fifty tons of silage. This amount will furnish silage for about ninety cows for eight months in the year. In case more than fifty cows are kept, two silos of medium size are preferable to one very large one.

Wooden silos are made either by bending boards around a cylindrical form or of staves grooved on the sides and fitted together like the staves of a barrel except that there is no bulge. The staves are held together by great iron bands placed on the outside and screwed up in order to make the walls of the silo practically air-tight.

The Foundation

A stone or cement foundation is desirable for all types and the bottom of the silo may extend from three to five feet underground in order to increase the depth of the silo without adding to its height above ground. The chief objection to having it extend too deep is the difficulty of getting the silage out when it is desired for feed. In ordinary practice the silo is built on the outside of one corner of the barn and a door opens from the silo into the feeding room of the barn or sometimes into the cow stable directly. Silos are also erected in the pastures in case silage is to be fed during the middle of the day or to beef cattle which are not brought into the barn for milking.

CHAPTER VI

SOLVING THE HELP PROBLEM

THOUGH dairying is one of the most profitable lines of farming, one must not lose sight of the fact that it involves a great deal of hard and confining work which must be done by some one. In certain parts of Europe, women perform a great deal of work connected with the dairy, milking particularly. Their services are cheap and efficient; in fact women are said to exert, through their personality, a beneficial effect on the milk-producing capacities of cows, a quality which the sterner and rougher sex does not possess.

In the United States, male help is used almost exclusively, and male help must be secured if a large-sized dairy is to be conducted. One man is able to care for from eight to ten cows, each giving a liberal flow of milk. A good milker can in average practice milk about eight cows in an hour, the production of the cows averaging about twelve pounds at a milking. A dairy of from ten to twenty cows in milk requires the services of two men to do the work expeditiously and for every extra ten cows an additional man will be required. To be sure, some dairy farms are conducted with less help in proportion to the number of cows, but in any short-handed business enterprise the work is either poorly done or the relations between the employer and hired help are constantly strained, due to the necessity of overtime work to care for the numerous details which constantly present themselves.

Make the Wage Scale Conform to the Work Scale

The help problem on a dairy farm is a difficult one, but it can be solved in the same manner that it is solved in city manufacturing establishments, namely, by paying high enough wages to attract good men, and treating the men so well that they will stay. The writer has little patience with those who deplore the scarcity of agricultural help and the absolute inability to secure competent help, while offering at the same time wages that are almost insulting. A decade ago it was possible to secure competent help for fifteen dollars a month in the South and about twenty dollars in the North. The prevailing wages now are about double those figures, but modern dairy practices are also more exacting in the thoroughness of the work. A competent herdsman or farm manager commands a salary of at least eighty dollars a month, and his board and room or house rent. Washing is not infrequently included when a large number of men are hired.

Diversion for the Hired Men

The nature of the work demands, furthermore, a higher scale of wages than is paid ordinary farm help. The hours are considerably longer on a yearly average and to many men work on a dairy farm appears to be a monotonous repetition of the same tasks. From the standpoint of hired help, there is an element of truth in this view which cannot be overlooked with impunity. Some compensating factor in the form of diversion must be supplied. The most satisfactory of all is a considerable number of hours set aside on a certain day each week for the hired man to use as he sees fit. This method, when there are a considerable number of milkers, permits of each man missing a milking occasionally and spending a golden half day in town or in the pursuit of pleasure elsewhere. If for any reason the above plan is not feasible, a long noon hour, from twelve to two o'clock, may operate to advantage, particularly if milking is begun early in the morning and is not over in the evening until a late hour.

If the best returns possible are to be expected from the herd, the intervals between milkings must be of about equal length. The best hours to begin milking are five o'clock in the morning and five o'clock in the afternoon; but this schedule cannot always be complied with. In the production of milk for the city trade, a certain train must usually be met, which necessitates early milking; but whatever the necessary hours are, they should be rigidly adhered to. Uniformity in the hours of milking must be impressed upon the men.

Another matter which hired men frequently dislike is miscellaneous work which was not mentioned at the time they were hired. There should always be a definite understanding as to the duties to be performed when the help is engaged. A man who engages himself as a milker or a feeder naturally dislikes to haul manure or peddle milk in town. These are merely typical instances indicative of points on which employer and employee may disagree where the duties expected of the hired man were not previously determined by the one and accepted by the other.

Both the hired man and the employer have rights, and the rights of each must be considered if the hired help problem is to be satisfactorily solved. The obligations on each side are about as follows:

Employer's Obligations

To state definitely the nature of the work expected, length of hours and wages to be paid.

To specify a definite person or "boss" from whom directions are to be received.

To pay wages promptly and in full according to prearranged agreement.

To allow his help a reasonable amount of time for recreation and for attending to his individual interests.

To recognize efficient service by an occasional increase of wages.

To provide for clean and comfortable sleeping quarters and good meals.

To treat his men with consideration at all times and when overtime work is necessary to pay extra wages for it.

Employee's Obligations

To furnish satisfactory references as to character, competence and industry and in no case to overstate his qualifications for the position he seeks.

To render conscientious services and work for the interests of his employer.

To act at all times the same as when the employer is at hand, particularly in the dairy barn and in handling the cows.

To recognize that, during the time specified, his time belongs to his employer and that the abuse of this time is equivalent to financially defrauding the employer.

To discuss directly with the employer and not with other hired help all sources or reasons for dissatisfaction or complaints.

To treat his employer with respect at all times and to do, when the occasion arises, any extra work which is obviously necessary, without being told.

To give reasonable notice of intention to leave or inability to be at hand at any time to assume his usual work or responsibilities.

The best manner of securing hired help varies with local conditions and with individual cases.

The most satisfactory results are generally obtained through personal solicitations among acquaintances in the dairy business. Another method is by corresponding with those in charge of agricultural schools or colleges where a large number of men fit themselves for dairy work. Still another means is by advertising in local newspapers or publications having a circulation in rural communities, such as dairy and general farm papers. Those who have been in the dairy business for a long time find it easier and more satisfactory to induce their men to stay by treating and paying them well, than to be breaking in new men constantly and having them leave through dissatisfaction shortly after they have become thoroughly familiar with the work and are just beginning to render efficient services.

CHAPTER VII

POINTS ON MILKING

THE relatively simple operation of milking is accompanied by such important secondary operations that a general consideration of the science of milking will greatly aid in securing the largest possible milk flow. The udder of a cow is spongy in structure and even just before milking contains a very small amount of milk; not more than a pint and a half.

During the operation of milking, the milk which partly fills the sponge-like recesses of the udder is drawn off through the teats and the milk-secreting glands immediately yield enough more milk to replace that withdrawn. As milking proceeds this process is rapidly repeated until all of the milk fluid tributary to the glands is exhausted. Milking is therefore an internal as well as an external process.

To properly perform it, place a low stool at the right hand side of the cow just in front of the udder. Grasp alternate teats firmly and draw the forefinger of each hand tightly around the teat just where it is joined to the udder. Then draw up the second finger, then the third and finally the little finger. This manner of applying pressure will force the milk in the teats out into the pail which is held between the knees. Always milk from the upper part of the teat down; never merely squeeze the teat. An "easy milker" is the term applied to a cow which will give down her milk freely and very little difficulty is experienced in milking such a cow. But when the milk fails to come after the teats have been squeezed several times or comes very slowly, the udder will require manipulation. This consists in kneading the udder by massaging it with the hands or, as more commonly practiced, by pulling down on the teats while the latter are being squeezed.

Importance of Rapid Milking

The mild stimulus thus given to the milk-secreting glands will start the flow and milking should be continued vigorously and rapidly until all of the milk is drawn. Rapid milking results in a more vigorous stimulation of the udder, its nerves and glands with the consequent large flow of milk. When the operation of milking is carried on slowly and listlessly, the amount of milk is quickly reduced and the profits of the dairy thereby lessened. To leave a cow partly milked to attend to some other duty, even for a minute, is a practice to be avoided.

"Stripping" is the term applied to the last part of the milking operation. After two teats have been thoroughly milked, the other alternate teats should be manipulated as described until all of the milk is apparently drawn. Then return to the first two teats and by drawing the forefinger and thumb of each hand down the entire length of the teats, a considerable amount of milk will be obtained. Repeat until all of this so-called "stripping milk" is drawn and then return to the second pair of teats, which are stripped in like manner. The stripping milk is very much richer than the first milk and all of it should be drawn. To neglect to strip a cow or to milk her only partially tends to dry her off, as is done a few weeks before calving.

In order to make milking a pleasure rather than a task, the dairyman needs but to take precautions to exclude flies from the stable, to see that the cows are comfortable and in harmony with their surroundings. A stranger in the barn, a recently whitewashed wall or a barking dog in the barnyard will in many cases make the cows nervous and irritable and difficult to milk.

CHAPTER VIII

PRACTICABILITY OF MILKING MACHINES

For over a century, inventors throughout the world have been endeavoring to make a machine which will successfully milk cows. Up to the present time the hundreds of patented inventions have given only partial satisfaction. The universal adoption of a milking machine will quickly follow the invention of one approaching perfection but, in order to reach perfection, the machine must milk more quickly and more cheaply and do the work fully as well as hand milking. The evolution of the milking machine will doubtless be similar to that of the cream separator. The first separators were large and bulky machines which required an engine to operate them. The modern farm separator is readily turned by hand and yields more cream with less power than did the earlier makes.

The "Sticking Point" in Machine Milking

The milking machines now in most general use

are those which require a considerable amount of machinery, piping and elaborate equipment in the way of tubes and specially made pails. Even these machines, with all their accessories and

X.—MILKING MACHINE PAIL with one set of teat cups in operation. The pail is divided by a partition which keeps the milk of two cows separately.

under skilful management, have not given complete satisfaction, although they have found favor in the eyes of some. Machines have been made to do seemingly more complex work than milking a cow, but such factors as the individual temperament and the physical structure of a cow are still baffling the inventors. These factors, together with the demand for machines which can be easily cleaned, makes the problem unique and therein lies the difficulty of making a perfectly satisfactory milking machine which can be used for all cows and managed by the average hired help.

Experience has shown that, in order to secure the best results with the machines now on the market, practically a new type of cow and an absolutely uniform type of udder and teats must be developed by breeding animals which have shown themselves to be adapted to mechanical or machine milking.

Proper Management Essential in All Cases

In addition to the power machines, there are a number of semi-successful hand and foot machines which can be placed between two cows and, by being operated with levers or cranks, will by suction, compression, or both, extract most of the milk from both cows. The work of these machines must generally be supplemented by hand milking in order to get all of the milk out of the udders. The experiences of those who have installed milking machines of various types vary so greatly that we are forced to the conclusion that the success of any milking machine depends ultimately on its proper management.

However perfect may be any milking device that can be invented, it will never be successful in the hands of the careless dairyman. Limiting ourselves to the power machines, which have been the most successful, we find that they consist of first, an engine of two horse-power or upward, a steel vacuum tank from which air is exhausted by suction pumps, air-tight piping which connects the vacuum tank with stanchion cocks near each animal, special pails to receive the milk and teat cups and tubes which connect the cows' udders with the vacuum pails.

The Mechanism of a Power Milking Machine

The operation of the milking machines is relatively simple. The accompanying diagram indi-

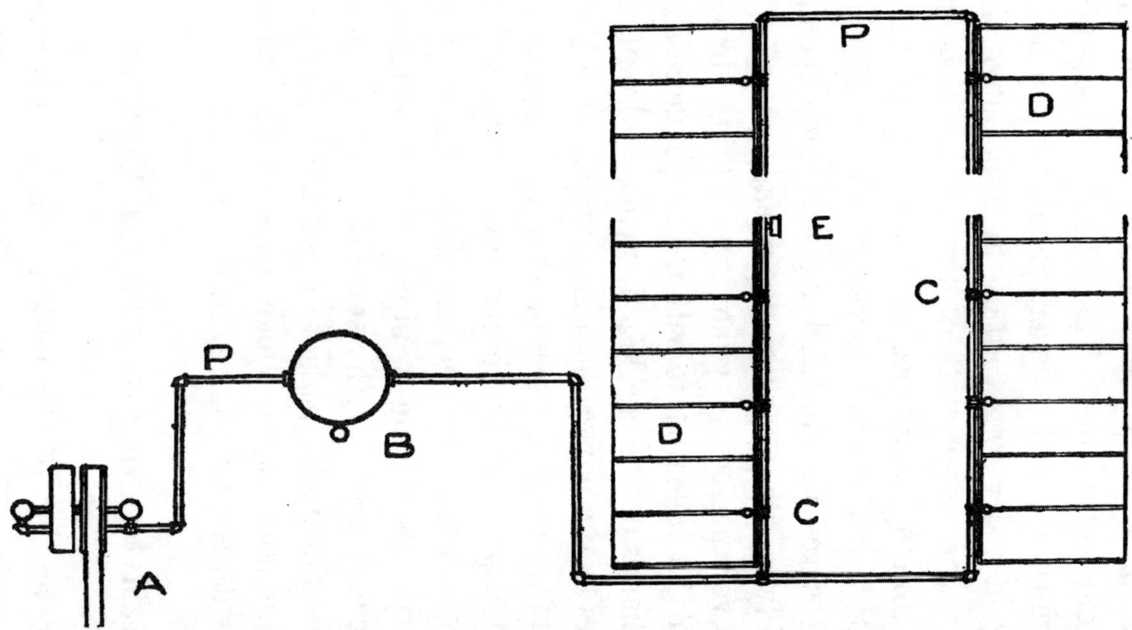

XI.—PLAN FOR THE INSTALLATION OF MILKING MACHINE SYSTEM. A, Vacuum pump. B, Vacuum tank. C, Stanchion cocks. D, Stalls. E, Vacuum indicator. P, Piping.

cates the arrangement of the machinery. The engine operates the vacuum pump at A, which gradually exhausts the air in the tank B. The vacuum reached and maintained is about fifteen inches, as indicated by gauges along the pipes and within sight of the milking machine operators. The vacuum pails are placed between the cows and the teat cups attached to the udders by simply placing them over the teats. They are held in place partly by the elasticity of the rubber lining and partly through suction. Modern machines have a pulsating device on the pail which creates an intermittent suction similar to the natural action of the calf in sucking milk from its mother. When all of the milk has been drawn from the cow, the operator notices, through a glass-covered opening in the pail, that no more milk is entering. The teat cups are then removed. During the process of milking, rubbing or kneading the udder will often stimulate a large milk flow.

Very few cows seem to object to being milked by machine, in fact some are more quiet when machine-milked than when milked by hand, due undoubtedly to the more uniform rhythm and gentler action of the machine than of the milker's hand. The greatest difficulty experienced and reported by dairymen has been the holding up of milk by certain cows. In milking a cow by hand, the milker is familiar with the temperament of the animal; he knows whether the cow is giving down her milk freely, and in case she does not, he supplies more power or milks more vigorously. Some teats are also more difficult to draw milk from than others and the hand milker can adapt his manner of milking to obtain the most satisfactory results from all quarters of the udder. The milking machine, on the other hand, is unable to cope with such a situation, owing to the inability of a machine to adapt itself to conditions it is unable to detect.

Difficulty of Cleaning

Great as are the above-mentioned problems, perhaps the most serious of all objections to the milking machine lies in the rubber tubes through which the milk must pass. Rubber cannot be sterilized by steam or high degrees of heat without resulting in rapid deterioration, which would in turn involve great expense if the rubber tubes had to be replaced frequently. Therefore, the cleaning of the tubes must be accomplished by careful washing. With such a perishable product

XII.—CLEANING MILKING MACHINE CONNECTIONS.

as milk, washing can never be too thorough, and to remove all of the milk in the tubes to prevent contamination of the next milking is a difficult task. Disinfectants or any of the ordinary chemical poisons which would sterilize the interior walls of the tubes cannot be used, owing to the serious danger of their affecting the purity of the milk later to be drawn through the tubes.

The only cleansing material which can be safely used is hot water, which certainly cannot be relied upon to remove all of the milk. Since the milky residue which is bound to accumulate in time on the inside of the rubber tubes cannot be seen, it is very likely to be overlooked and its presence neglected. This is the master objection to the present milking machines.

Saving of Labor is Problematical

The saving of help is a question which can be answered in neither the negative nor the affirmative. True, one man can attend to several machines and at certain times can be milking as many as twelve cows simultaneously, but at other times, when he is getting the machines ready, he is not milking any cows. On large dairy farms, where fifty or more cows are milked regularly and where there are several milkers, the installation of a power milking machine system might result in the saving of the wages of a man or two. In such a case we can say that the milking machine is a labor saver. But on a small dairy farm, where less than fifty cows are kept, the advantages of a milking machine, as a labor saver, are problematical. Certainly, where efficient help can be secured at a reasonable scale of wages, the milking machine is not advisable. When help cannot be secured readily, the machine may perhaps solve that problem.

The initial cost of a power milking machine system for a herd of forty cows is in the neighborhood of five hundred dollars with an additional twenty-five dollars a year for upkeep. This does not include the cost of the power required to operate the pumps. The effect of milking machines on the health of the herd is beneficial rather than detrimental, owing to the somewhat better care which is taken of the cows' udders where machines are used.

CHAPTER IX

FEEDING AND MANAGING THE HERD

THE feeding of dairy cows differs from the feeding of horses, dogs and the various domestic animals only in the respect that the cow responds more quickly to changes of feed than any other animal. This response is particularly noticeable at the pail and by watching carefully the amount of milk produced each day, the skilful herdsman is able to vary the ration so that each animal will produce her milk most liberally and economically. The composition of the animal body is very similar to that of the human body, and such foods as fats, proteids and those rich in starch are as necessary for the dairy cow as for the human system.

The chief difference between the feed which a cow can consume and assimilate and that which a human being eats, is that the cow's food may be much coarser. The cow can convert crude fats and oils, such as are found in cottonseed meal, flax and to some extent in all grains, into the very valuable product known as milkfat or butterfat. She can convert the proteids of hay, grass, and corn silage into the delicate albumen found in milk. The value of a dairy cow depends on the readiness with which these changes take place in her body and also to the ratio between the amount of food consumed and the amount of milk produced.

Succulent Feeds and Pure Water

As previously stated, in the discussion of breeds, the difference between a dairy cow and a beef or dual-purpose animal is that, in the former case, the feed is converted almost entirely into milk and very little is taken out by the cow to be stored up as fat and tissue. In the latter cases, where the feed is converted into flesh, which deteriorates in market value as the animal becomes older and yields no profit at all until the animal is killed, the cost of the feed is so great in proportion to the amount of milk produced, that the animals cannot be kept at a profit except where there is extensive pasturage or where the feed is very cheaply obtained. As water constitutes about eighty-seven per cent. of normal milk, the importance of succulent feeds and of a liberal supply of pure water cannot be overlooked.

The Balanced Ration

One is not very long in the business of feeding farm animals, particularly dairy cows, before he hears of the balanced ration. By a balanced ration is meant a plan of feeding, so that the percentage of digestible matter in all of the foods totals the body requirement of the cow. Just as meat, potatoes, bread, butter and pudding are said to make a well-balanced meal for the human family, so a definite ratio of roughage, concentrates and occasional condiments make a balanced ration for the dairy cow.

No single standard of feeding can be laid down for dairymen, since the standard constantly changes with the climate, temperature and the condition of the animals at different times of their milk flow. As a tentative guide, the following figures represent the ration of a cow weighing about one thousand pounds and giving about ten pounds of milk twice a day: two pounds of proteid food, twelve pounds of carbohydrates and one-half a pound of fat.

Variety of Seasonal Rations

Of course a very large variety of foods can be drawn upon in making the necessary selection at any time of the year. In the spring months the common roughage feeds are as follows: timothy hay, red clover hay, alfalfa hay, wheat straw, corn silage and sometimes dried corn stalks. The concentrates for the spring months are generally soy beans, cow-peas, oats, cottonseed, and barley meal.

For summer feeding the roughage is generally pasture grass and clovers, hay and some silage, depending whether or not the latter is at hand. Concentrates for the summer months are fed sparingly and should contain a rather small percentage of fats and oils. Such mill products as rye flour, ground corn and oats, and corn-cob meal are typical of the kinds of feed which may be fed. Waste products or by-products of other industries such as brewer's grains, barley screenings, and cottonseed hulls may also be fed.

In the fall months the roughage may be fall pasture, sheaf oats, fresh corn and the season's hay. The concentrates should be about the same as those fed during the summer.

Palatable Winter Feeds

The winter months are those which tax the skill of the feeder to the utmost in securing a large flow of milk. Milk and cream prices are highest during the cold weather, and to produce milk cheaply for a high-priced market is naturally the end sought. Dairymen all over the country are recognizing the importance of laying by a great variety of palatable feeds for winter use. Among the roughage feeds are corn silage, alfalfa hay cured in the cock, cow-pea vine silage, kafir corn stover,[1] and red clover silage.

Among the concentrated foods for winter feeding may be mentioned oil cake, cottonseed meal, flax, gluten meal and all of the grains. Especially important in the winter are roots and tubers. The English and Danes feed more root crops, as they are called, than the Americans. The great advantage of root crops are that they are in a way condimental and are instrumental in keeping the system in good order. Among those which may be laid by to advantage for winter use are potatoes, both Irish and sweet, sugar beets, turnips, rutabagas and carrots.

[1] Kafir corn fed in its natural state, with the heads left on. "Stover" means "fed whole."

Natural Feeds Safer than Proprietary Preparations

The above-mentioned rations are suitable for milch cows. Growing calves require less roughage and more concentrated food. Calves thrive especially well on skim-milk in which there is ground flaxseed, and various meals derived by crushing grains. There is on the market each year an increasing number of prepared foods of concentrated nature for dairy animals. So great and so unfounded have been the claims of certain manufacturers of these preparations that many states have inaugurated an inspection of feeding stuffs in order to prevent worthless food from being sold at exorbitant prices.

Obviously a few tablespoonfuls of a preparation cannot result in a rapid gain in weight by the animal, neither can it increase the milk flow permanently. A great many of the preparations are little more than stimulants and frequently injure the cow's delicate digestive and milk-producing mechanism. One is always following the safest line of feeding when he limits his rations to natural feed. To be sure, there is an attractive mystery connected with feeds brought in at great expense, particularly when the feed is accompanied by strong claims for its merits.

The cheapest and the best foods, however, are usually those produced near by and, in case one desires to feed manufactured preparations, he should write to his state experiment station for information as to the real merits and feeding value of the feed he proposes to use.

Cattle Feed is Expensive Fuel

A very important consideration in winter feeding is the warmth of the stables. A dairy cow, producing a large flow of milk, is a sensitive creature and should be adequately sheltered both in the barn-yard and in the stable. A stable, free from draughts and a barn-yard protected on one side at least from bleak winds, will more than pay for themselves in the course of time by the increased amount of milk produced by the herd. In very cold weather the drinking water should be warmed slightly in order that a large amount of feed will not be consumed for raising the temperature to that of the cow's body. The reader can see the economy of using fuel as the source of heat for warming the water rather than drawing upon the vitality of the cow.

During the winter months, feed rich in fat and protein should be used more freely than during the summer months in order to make the cows more resistant to cold. The general rule for feeding a dairy cow grain (which is relatively rich in fat and protein) is to feed her as many pounds of grain a day as she produces pounds of butterfat in a week. Stated another way, she may be fed about one-fourth as much grain per day as she produces pounds of milk. During the months of cold weather, the careful herd manager will give the cows, which are producing large amounts of milk and butterfat, a larger ration of grain than those producing smaller amounts.

The Nutritive Ratio

The nutritive ratio of feeds is the ratio between the amount of digestible protein in the feed and the combined amount of digestible carbohydrates and fats. For example, the nutritive ratio of such a product as wheat bran which contains about four parts of digestible carbohydrates and fat to one part digestible protein is 1 : 4. A ratio in which the last number is not more than four times that of the first is a "narrow" ratio, while a ratio in which the last number is more than six times that of the first is known as a "wide" ratio.

The nutritive ratios of a few of the most common concentrated feeding stuffs are as follows:

Cottonseed meal,	1 : 1.2	Middlings,	1 : 4.2
Oil meal,	1 : 1.7	Ground corn and oats,	1 : 9.0
Wheat bran,	1 : 4.1	Hominy,	1 : 11.4

Fat Content of Milk not Affected by Feed

As a rule the feeds rich in protein are slightly greater in cost than those with a wider nutritive ratio. From the large number of feeds with different analyses and widely varying percentages of fat, one might think it a simple matter to select feeds that would make the milk very rich in butterfat. Scientists have worked for years on this problem, but without practical success. There is no selection of feed which will cause the percentage of butterfat in the milk to increase. The only means of artificially affecting the fat content of milk is to almost starve the animal. In this way the amount of milk is very much reduced, and as the cow draws on the fats stored up in her body for the fatty material in her milk, a slight advance in butterfat over the normal content may be noted. Such a gain is entirely offset, however, by the corresponding loss when the animal is again fed a full ration. Obviously the method above mentioned has no commercial value.

The management of the dairy herd simply demands good business methods, and careful watching of the disposition and performances of each animal is practically all that is necessary as a guide to proper care. Dairy cows will, of course, do best when they are kindly handled and milked. An animal which has been forced into her place by the use of a pitchfork and curried down with the leg of a milking stool, is not ready to do justice to her normal ability to produce milk. The cows should be treated like ladies, and when there is a possibility that an animal does not understand what is expected of her, she should always be given the benefit of the doubt.

Management of Bulls

An unruly or a fractious cow should, of course, be treated firmly but not roughly. The same rule applies in the case of bulls. Males of all of the dairy breeds should be handled as though they were dangerous, however gentle they may at times appear.

Owing to the great strength of a bull, a jockey stick snapped to a copper ring in the nose is the safest way to handle him. If always kept completely under control, he will not become aware of his strength, whereas should he be tied or led

XIII.—GUERNSEY BULL BEING HANDLED WITH JOCKEY STICK. The snap at the end of the stick catches into the copper ring in the bull's nose and he can neither run toward nor away from the person handling him.

about with merely a rope and halter and should succeed in breaking it once, the wisdom he thus gains may be put to wicked use when he is least suspected.

CHAPTER X

TESTING MILK FOR RICHNESS AND PURITY

THE considerable number of municipal, state and federal standards which dairymen must comply with in order to put out legal dairy products, necessitates a thorough understanding of the principal tests applied to dairy products. Among the requirements which are most prominent in specifying the purity of milk are those which define the lowest percentage of butterfat in the milk, the lowest percentage of total solids in the milk and those which further specify that nothing can be added to or subtracted from milk to be sold as whole milk. The term whole milk means milk as it is drawn from the cow with no cream added or removed.

If one is putting out certified milk, the matter of the bacterial count is of importance. The principal standards for cream are purity and richness. Most of the standards for butter specify that it must contain a minimum fat content of eighty per cent., although a few states still adhere to an unreasonable standard of eighty-two and one-half per cent. Ice-cream standards specify particularly the lowest permissible amount of butterfat.

The chief federal standard, which is very strictly enforced, is that which deals with the percentage of moisture in butter. A ruling of the Bureau of Internal Revenue specifies that butter containing sixteen or more per cent. of water is adulterated, and that the manufacturer of such butter must pay a tax of ten cents per pound and, in addition, take out a license for manufacturing adulterated butter.

The Fraud of One-Minute Churns

This ruling bears especially on the products of the so-called one-minute churns, recently placed on sale, and advertised widely among dairymen. The manufacturers and agents of these churns claim that a pound or a pint of milk, mixed with a pound of butter, will produce two pounds. The inference is that the product is butter, but in reality it is always adulterated butter. The standards of purity for milk products, as determined by the Secretary of Agriculture, are not strictly enforced, but have served as valuable guides for the enactment of stringent state and municipal laws. As these laws are constantly changing and vary in different localities, every dairyman should secure a copy of them from his state dairy commissioner.

Operation of the Babcock Test

The test most widely used among dairymen is the Babcock test, named after Dr. Babcock of the Wisconsin Experiment Station. This test enables one to very quickly, accurately and cheaply determine the percentage of butterfat in milk, cream, ice-cream and cheese. It consists of measuring, in the case of milk, and weighing in the case of the other products a definite sample which is placed in a graduated test bottle. About an equal amount of sulphuric acid is then added and the mixture is whirled for a few minutes in a centrifuge. Enough water is then added to bring the fat, which has been separated from the rest of the mixture, into the neck of the bottle and the richness of the milk or other product is read directly by means of the graduations of the neck of the bottle. A separate bottle is required for each class of products.

These testing outfits may be purchased for as

little as four dollars, and every dairyman should, by all means, own and use a Babcock test for milk at least. A test can be made for about one cent and the time required is not more than ten minutes. The directions which accompany different makes of the testing outfits are always clear and easily followed. In view of the wide difference of makes of testing bottles and slight differences in the methods of handling the various makes of tests, no general directions can here be given.

The Test for Solids in Milk

The tests for the total solids in milk are seldom made by dairymen who are putting out a high grade of milk. The standards are sufficiently reasonable to permit the average milk of any herd to pass inspection, but in case watering is resorted to (always in violation of the law) the offender is usually caught and forced to pay the penalty in addition to receiving undesirable publicity.

The common test for the total solids in the milk has for its basis the use of a lactometer, an instrument which floats in the milk. It has a graduated neck and the denser the milk the higher the lactometer will float.

Test for Moisture in Butter

The moisture content of butter is determined by first weighing a definite amount of the butter and then heating it sufficiently to drive off the moisture. When the moisture has been entirely expelled, the sample is again weighed and from the loss in weight the percentage of water is calculated. While this test is used chiefly in creameries, a dairy farmer, putting a large amount of butter on the market, can use one of the many makes of moisture tests to great advantage. Butter will very seldom contain more than sixteen per cent. of moisture unless special means and methods have been followed to incorporate a large amount of moisture in the butter. On the other hand, a great deal of dairy butter contains an unusually small amount of moisture. A few per cent. more would not only increase the profits of the dairyman, but would, in many cases, improve the quality of the butter. It is obvious that the use of a moisture test will quickly pay for it if the conditions just mentioned prevail.

The Hart Casein Test

Another test, invented by Dr. E. B. Hart of the Wisconsin Experiment Station, is one for quickly and accurately determining the percentage of casein in milk. This test is now on the market and is being used by many cheese factories as the basis or at least a guide for paying farmers for milk delivered to cheese factories. Its operation is somewhat similar to that of the Babcock test and each of these casein tests sold is accompanied by full directions.

The Milk Sediment Test

One of the most striking tests used in the dairy is the milk sediment test. The tester consists of a cylinder surrounded by a hot-water jacket and provided at its lower tapering end with a cap containing a circle of wire gauze. A bit of absorbent cotton is placed over the gauze and milk is poured into the cylinder. After the milk has run through the cotton and the wire gauze, the clamp which holds them in place is removed and the cotton disc is examined for sediment. If the milk is clean, the cotton disc will be perfectly free from a dark coloration; but if dirty milk is put through the tester, a black circle will be left on the cotton. Varying shades of gray on the cotton indicate the relative purity of the milk from the standpoint of dirt.

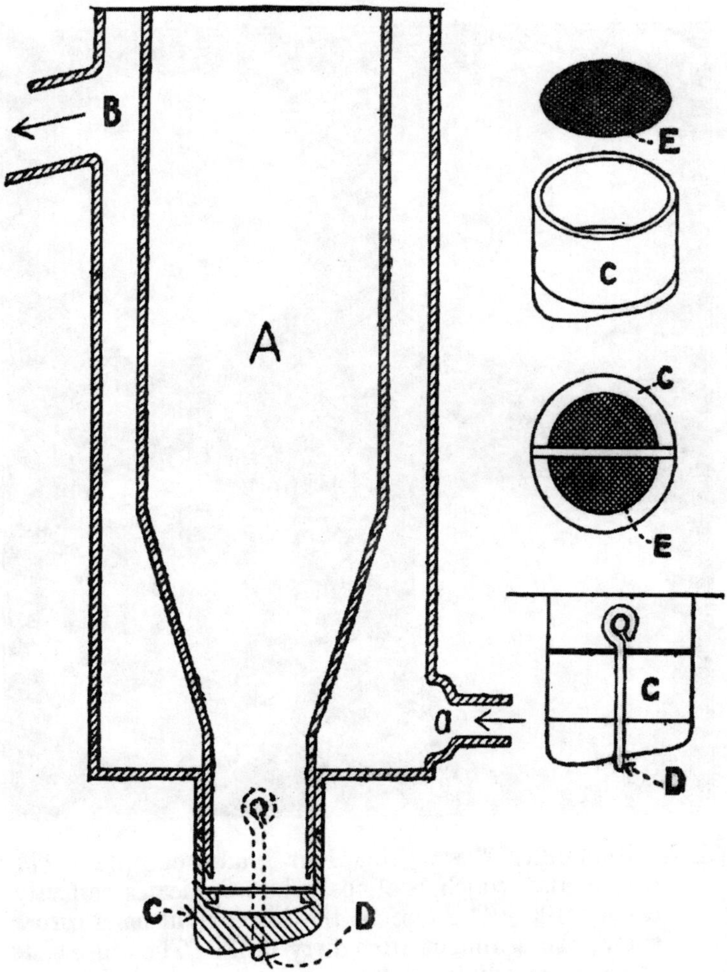

XIV.—MILK SEDIMENT TESTER. A is the central cylinder holding about a pint. B is a hot water jacket with receiving and overflow apertures. C is a brass cap containing a circle of wire gauze E, over which is placed a disc of absorbent cotton. D is a clamp rod for holding the cap and cotton in place.

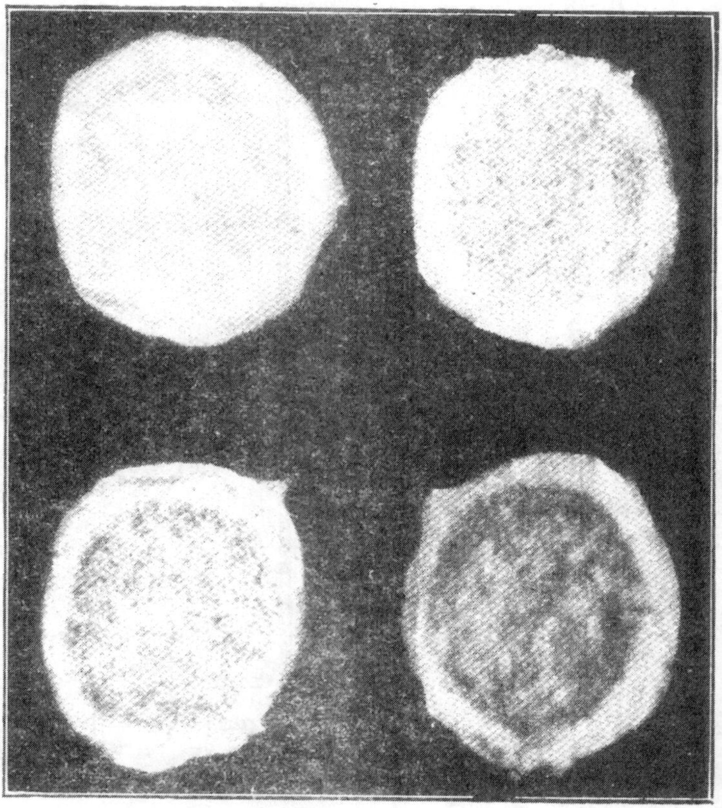

XV.—SEDIMENT TESTS from four grades of milk. The cotton disc, which is almost white, indicates perfectly clean milk. The disc in the lower right hand corner shows the sediment from dirty milk. The other tests show intermediate grades of milk.

TESTS FOR BACTERIA

Tests for the number of bacteria in milk must be conducted by a skilful bacteriologist if accurate results are expected. The number of bacteria found in a unit amount of milk, generally a cubic centimeter, is influenced by the following factors in the history of the milk since it was drawn: cleanliness of milking, temperature at which milk has been held, cleanliness of bottles or other containers and the age of the milk.

Other tests which are commonly applied to milk, cream and butter are those relating to the acidity, or relative sweetness, and in the case of butter the test for salt. These tests are not commonly applied in the dairy but their existence is mentioned so that they may be obtained, if desired, from any manufacturer of dairy apparatus.

CHAPTER XI

SANITATION AND CARE OF MILK

THE same factors which make milk valuable as a food also make it one of the most perishable of all foods. These factors are: its delicate chemical nature, its liquid form and its liberal percentage of animal fat. Its delicate chemical nature makes it a favorite food medium for many kinds of bacteria, its liquid form causes any contamination to spread rapidly through the entire amount and its fat content renders it susceptible to taint from various undesirable farm and kitchen odors. The reader will remember that fatty substances are used in the extraction of perfumes from flowers, and this same principle applies in the opposite extreme to the contamination of milk and cream by objectionable odors.

Normal pure milk contains about 87 per cent. of water, 3.7 per cent. of fat, 4.8 per cent. milk sugar, 3.8 per cent. casein and the remaining .7 per cent. mineral matter. When milk leaves the cow's body it is practically sterile and free from taint. The care that is taken of the milk from that time until it is delivered to the consumer determines its value, its quality and its price.

Bacteria in Milk

Throughout the air, water and land of the earth are microscopic plants (not animals), invisible to the naked eye, but which rapidly multiply to produce visible results. About ninety-nine per cent. of these bacteria are beneficial so far as they destroy decaying organic matter, help the farmer to maintain the fertility of his fields and warn the public of dangerous foods, dwelling places and other menaces by the objectionable nature of the gases and odors they produce. The most common germ found in milk is the lactic acid bacillus or bacterium (plural, bacteria) which, if allowed to develop unchecked, causes the souring of milk. The souring process is merely the development of so much acid that the taste is unpleasant to the human palate. Throughout all lines of the dairy industry, one is constantly working with or against bacteria, chiefly the lactic acid bacteria. In butter and cheese making, the growth of bacteria is encouraged to a considerable degree in order to secure the desired bacterial flavors. In milk and cream, on the other hand, the dairyman must wage a constant fight against bacterial development in order that the milk and cream may stay sweet for a sufficiently long time to convince the consumer that the milk was pure.

Value of Heat and Cold as Dairy Agents

The best known means at the present time for controlling bacterial development are cold, heat and preservatives. The lactic acid bacteria make their best growth at a temperature between eighty and one hundred and twenty degrees. Above and below those limits the rapidity of growth becomes less as the temperature increases toward the boiling point and decreases toward the freezing point. The producer of milk relies, for the most part, on cold temperatures in order to retard the development of bacteria, and to keep the milk in a sweet condition as long as it is in his hands. With a cold temperature there must also be the exclusion of air, dusty air particularly. If dust is allowed to gain access to the milk, even though the latter is kept cold, no serious results will be visible for some time but, as soon as the milk approaches seventy degrees or above in temperature, a tremendous growth begins and the souring is completed in a short time.

Solving the Dust Problem

The simplest ways of guarding against dust are, first, to avoid feeding such dusty rations as hay or dry meal just before milking; second, to wipe the udders and teats of the cow with a damp cloth just previous to milking; third, to avoid wearing clothes which have been worn in the field or otherwise exposed to dirt; fourth, to use carefully cleaned and sanitary milking pails and cans. The most suitable pails for milking are those which are partly covered on the top yet allow sufficient space for the streams of milk from the teats to be directed into the pail without loss. The use of strainers in the dairy is a matter which is not well understood. Straining milk removes visible straw, manure and filth which find access to the milk even when the milking is apparently carefully done. The strainer, however, does not remove the millions of bacteria which were formerly lodged on this foreign matter, and they are always washed through the sieve of the strainer into the milk. Straining the milk obviously does not serve to purify it in any manner except to remove the visiible means of the contamination. In a well-regulated dairy, straining the milk is done only to remove the foam which is merely a small amount of milk blown up by air bubbles. A liberal amount of foam is always produced by rapid milking.

Where a strainer is used, it should be of the nature of a very fine wire screen and not cloth or material which cannot be thoroughly cleansed. This rule does not apply to the very excellent milk pails over the top of which is a layer of absorbent cotton used only once. The latter system for keeping foreign matter out of the milk is used chiefly in producing a very high class of certified milk, and is rather too expensive for adoption by dairymen in general.

Pasteurization and Sterilization

Having kept the milk cold, generally by placing it in well-covered cans which in turn are almost submerged in cold water, the dairyman producing milk for the market trade is able to supply the city dealer with a very creditable product. Heating, as a means for keeping the number of bacteria in the milk low, is commonly employed in two operations, pasteurization and sterilization. Pasteurization derives its name from the famous French bacteriologist, Louis Pasteur, who discovered the process. The name should not be confused with the English word "pasture."

Pasteurization merely consists of heating milk to temperatures ranging from 140 degrees Fahrenheit to 180 degrees Fahrenheit. This range of temperatures varies according to the length of time the milk is to be kept or pasteurized at any particular degree of heat. When a temperature of 180 degrees is used, the process is practically instantaneous while a temperature of 140 degrees must be maintained for about twenty minutes in order that the lactic acid bacteria may be destroyed. Various cities have ordinances or health board rulings which specify the temperature to be used and the length of time which the milk must be held at that temperature in order to be sold under the name of pasteurized milk. The effect of pasteurization upon the taste and digestibility of milk is a much-disputed problem. Pasteurization must not be confused with sterilization. The latter is the scalding of milk at within a few degrees of its boiling point. This process coagulates the albumen and the casein and decreases the digestibility of the milk. Pasteurization, if properly carried on, need not have any injurious effect upon the quality of the milk, and should under no circumstances give it a scorched taste. Pasteurization furthermore is the present chief safeguard against the transmittance of contagious diseases, including tuberculosis, through the milk supply.

Abuse of the Principle of Pasteurization

Pasteurization, with all its advantages and benefits, has two main drawbacks. The first and most important is that dealers and consumers of pasteurized milk abuse the element of safety which pasteurization provides. Owing to the fact that pasteurization reduces the tendency of milk to sour quickly, pasteurized milk is often kept in the household for several days. During this time other types of bacteria beside the lactic acid germ develop. Some of these germs do not produce acidity or sourness, but may cause the milk to undergo fermentation or peculiar forms of curdling and decay. Pasteurized milk, several days old, is therefore dangerous, not from any fault of the pasteurizing process, but because the public has abused the safeguard which pasteurization affords.

The second objection to pasteurized milk is the slowness with which the cream rises and the incompleteness of the separation. The heating process thickens the milk very slightly but enough to interfere with the rise of the fat globules, collectively known as cream. While this objection to pasteurized milk is unjustified, nevertheless the public likes to see a liberal amount of cream on the milk.

Chemical Preservatives are Unlawful

The use of preservatives will keep the milk sweet for an indefinite period of time if enough of the preservatives are used. Certain chemical compounds such as formaldehyde, boric acid and other almost tasteless preservatives are used by unscrupulous dairymen in order to cover up careless handling and to give their milk the desirable quality of remaining sweet a long time. Such practices are always against the spirit of the law and generally in strict violation of the letter of the law. Common table salt, a preservative allowed in butter where it also serves to bring out the flavor, is barred from milk and cream even when these products will subsequently be converted into butter. Dairymen should beware of all powders, liquids and preparations offered for sale for the purpose of lengthening the age of marketable milk. Cold, heat and cleanliness are the sole auxiliary forces which the dairyman can legally use in putting out a good and sanitary grade of milk.

CHAPTER XII

CERTIFIED MILK AND ITS PRODUCTION

THOUGH certified milk is commonly considered the highest and purest class of all market milk, the term "certified" has only a relative application. Certified milk came into public notice about the year 1897 as the result of crusades, especially in some of the larger cities, for better milk.

The value of certified milk depends on the standing of the certifier to its purity and what the standards of purity are. In most cities, responsible citizens interested in a pure milk supply form a commission, the officers of which are physicians, veterinarians, chemists and milk experts of high standing. Inspectors are appointed who visit the dairies producing the milk to be sold as "certified" and see that all standards and requirements are strictly complied with. The conditions under which milk is commonly classed as "certified" are as follows:

The herd must be examined at frequent intervals, not more than six months apart, by a reputable veterinarian and must prove to be free from all diseases.

Water must be pure, feeds free from taint or decay, pastures free from garlic or strong-flavored weeds, barns clean, milkers cleanly in person and habits, and milk taken care of according to rules established by the commission.

The rules generally specify that milk must be cooled to fifty degrees or less immediately after it is drawn and delivered to the consumer in the shortest possible time, ranging from a few hours up to twenty-four depending on transportation facilities.

Visits of Inspectors are not Announced

As a check on the producer, inspection of his premises and herd are made without announcement or warning and his milk is examined at frequent intervals for its bacterial content. The latter term means the number of bacteria per cubic centimeter or about a teaspoonful. Ordinary unpasteurized market milk contains upward of a million bacteria per cubic centimeter when delivered to the consumer and this number increases rapidly in the course of a day. The maximum bacterial limits established for certified milk are between 10,000 and 100,000 bacteria per cubic centimeter. It is impossible even with advanced scientific methods to put out large quantities of milk containing on an average less than 10,000 bacteria per cubic centimeter. On the other hand, milk containing over 100,000 bacteria per cubic centimeter cannot properly be classed as certified as such a high figure does not indicate careful methods of production.

Summer and Winter Standards

Some medical or certified milk commissions have wisely made two standards of bacterial count, one for the summer months and another for the winter. Recognizing the difficulty of maintaining the milk at a low temperature in hot weather, these commissions permit the bacterial count to be double that allowed in cold weather. The average number of bacteria per cubic centimeter in all certified milk runs close to 50,000 in summer and 25,000 in the winter time. Certified milk, while not absolutely pure and free from bacteria, approaches ideal milk more than any other form, but its high cost of production must be accompanied by a high retail price, and here lies a problem which the producer must solve with his own ingenuity.

Various Grades of Milk Compared

Until recent years milk has always been just "milk" and the public has not learned to recognize that pure milk is worth more than impure or infected milk. The producer must by judicious advertising and by an educational campaign demonstrate that the good milk he places on the market is cheaper, considering food value and health, than a poorer quality of milk, even though the latter is sold at a lower price. In the present commercial age, however, certified milk has not made very great headway. Pasteurized milk is perhaps the safest, and the milk to be most highly recommended for purchase by the consumer of moderate means. Pasteurized milk costs about twenty-five per cent. less than certified milk and about ten per cent. more than the cheapest form of raw milk.

The value of certified milk depends upon the rigidity with which the inspection of dairies is maintained. If the inspection becomes lax at any time, certified milk is of no more value than ordinary raw milk. Pasteurized milk, on the other hand, goes through the daily process of pasteurization, which is practically uniform at all times.

CHAPTER XIII

THE SCIENCE OF BOTTLED MILK

WHILE manufacturers of breakfast foods and canned goods learned at a very early stage in the development of their respective businesses the value of an attractive package, dairymen have been much slower in this respect. Butter was one of the first dairy products to be sold in artistically-designed wrappers, and now a great deal of it goes on the market in the form of attractive prints. With the exception of the fancy cheeses, cheese is still sliced off in the average grocery store and wrapped in common manilla wrapping-paper. This fact has been one reason for the very dull edge of the cheese appetite of the American people.

In the sale of milk to the city trade, the appearance of the package has much to do with the demand and to the size to which any dairyman's business may grow. Many cities have ordinances specifying that all milk containers must be kept inside of the distributing wagons and that milk bottles cannot be refilled before thoroughly washing and sterilizing them at the bottling plant. Other cities require all milk to be sold only in bottles. While most of these regulations have been established purely from a sanitary standpoint and for reasons bearing on the public health, they have also led the public to recognize the dangers from the use of indiscriminate milk receptacles and have promoted the present popularity of bottled milk.

The Glass Jar and the Single-Service Package

Though bottling adds to the cost of milk it also adds to its value and the public is willing to pay the price. However slow they may be to appreciate the value of pasteurized and certified milk, the American people know that bottled milk looks better to the eye than a bowl of milk which they have seen poured out of a battered tin receptacle. Many forms of milk bottles have been experimented with, but the thick glass jar has proven the kind best adapted to stand rough handling and the rather severe action of the mechanical washing and filling machines. The single-service bottle made of heavy paraffined pasteboard was widely exploited for a time; but it has not proven satisfactory, owing to one chief objection. The consumer is unable to see how much cream has risen to the top of the milk. Other objections are their unattractiveness, their expense and their inability to stand rough handling.

Different Types of Caps

The usual form of milk and cream jar is provided with a small rim cut inside of the neck to receive a paraffined paper cap. The chief objection to this kind of cap is the tendency for dirt to accumulate on the inside of the neck just around the paper cap and when the latter is removed and the milk poured out, there is always danger of contamination.

This objection has been overcome by the application to milk bottles of the same form of cap used on carbonated beverages, namely, one which completely covers the top of the bottle, thus protecting it completely. The cap is placed over the bottle and clamped on with rapid acting machines. As the inside of the metal cap is provided with a paraffined disc which exactly covers the top of the milk bottle, the contents are hermetically sealed; in fact the seal can be so tightly

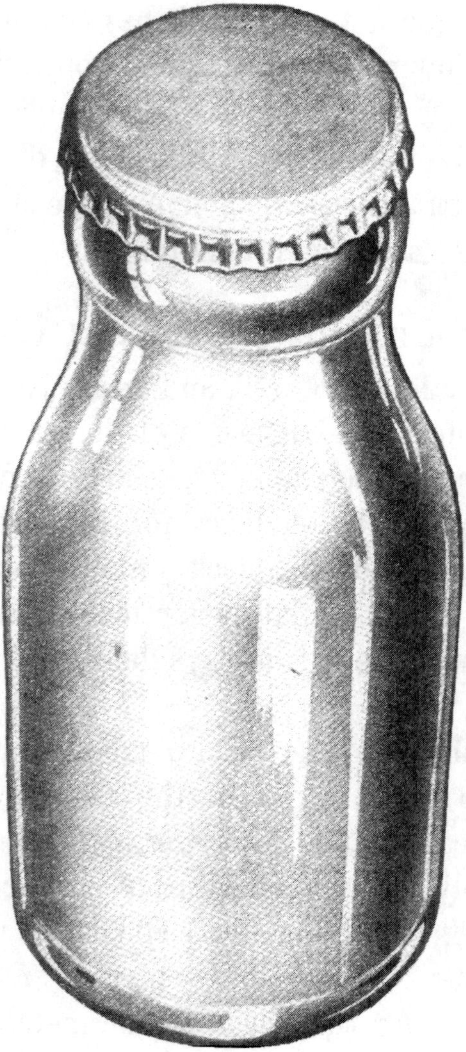

XVI.—HALF-PINT MILK BOTTLE with improved metal cap. There is no chance for dirt to lodge on the top of the bottle and the danger of the bottles being chipped is reduced to a minimum.

applied that the milk can be pasteurized in the bottle and the only possible chance for contamination is after the crown cap has been removed by the consumer.

The Package as an Index of its Contents

Another advantage of this system is the impossibility of bottles being refilled with low grade milk without detection by the consumer, since when the cap has once been removed it can never be made to fit the bottle tightly again. While the metal caps or crowns, as they are commonly called, are slightly more expensive than the cheap paper caps, the advantages of the system outlined appeal to the best trade, with which it is fast finding favor.

The highest grade of package so often accompanies the highest grade article that dairymen should take advantage of this knowledge and sell their product in a form which will inspire the confidence of the purchaser. Just as a banker cannot gain the confidence of the public if he dresses in overalls and jumper, so even the purest milk cannot be expected to be in demand if it is sold in untidy bottles.

CHAPTER XIV

MILK INSPECTION AND ITS PURPOSE

THE dairyman or the person going into the dairy business must expect to be intimately associated with dairy inspectors at frequent intervals. The development of civilization has made necessary the enactment of certain reasonable regulations by which the public may be protected against the greed or dishonesty of unscrupulous milk dealers. As delicate tests are required to determine whether or not milk has been adulterated, the public is unable to determine for itself whether its milk supply is pure. For this reason its elected representatives have made laws and appointed experts who are constantly watching, if they do their duty, for unscrupulous practices among producers and dealers of dairy products.

Rigid dairy inspection makes possible the development of a healthy and prosperous dairy industry, since it eliminates dishonest practices and destroys dishonest competition. The inspector should be looked upon by a law-abiding citizen with as much respect and confidence as is due a sheriff, a police officer or other public servant charged with protecting the public's welfare. Milk inspectors also act as instructors, and even though their directions seem to impose hardships upon the producer, beneficial results are in the end almost invariably obtained. By reason of their wide experience and constant association with progressive dairymen, their advice should be heeded.

Rights of Producers, Inspectors and Consumers

Inspectors are vested with authority to inspect the premises of the dairyman at all reasonable hours of the day and to examine the herd, the barns, milk houses, and milk receptacles. Such intrusion on one's private property may appear uncalled for and be resented, especially by the dairyman conscientiously conducting his business. The consumer, however, has rights which must be protected, and the inspector has no way of determining who is doing an honest business or who a dishonest one without personally inspecting the premises at frequent intervals.

Were it not for milk inspection, the dairy business would be overrun with slovenly and dishonest people who would take advantage of the public confidence in good milk by selling adulterated and inferior milk at the same price. The states where dairy inspection is most rigid are those which have the most prosperous and most intelligent dairymen.

CHAPTER XV

DAIRY BUTTER—HOW TO MAKE THE BEST

In all of the large produce markets, dairy butter is quoted several cents less than that made in creameries. The commercial inferiority of butter made on farms to that made in creameries is due first to a lack of uniformity and second to a great deal of poor dairy or country butter. Its inferiority is due to a considerable degree to unskilled methods of making, careless packing and improper care after it is made. Frequently traded off at the country store for groceries and supplies, it is held at high temperatures until it becomes rancid and the only disposition which can be made of it is to renovating factories where the bad odors are driven off by forcing a blast of air through the melted mass, which is then rechurned.

Owing to the small amount of butter which individual farmers are able to produce, they cannot take advantage of refrigerator freight cars and however good the butter may be when it leaves the farm, its quality suffers so severely in transit that its market value is quite low. Bearing in mind the above facts and the difficulty of disposing of butter made on the farm through the large produce markets to good advantage, let us see what advantages the farm dairy has over the creamery.

The Opening Era of Better Dairy Butter

The dairyman who desires to engage in farm butter making has absolute control of the source and the purity of his milk and cream. A good quality of finished product invariably depends on a satisfactory grade of raw material with which to work. Furthermore, he is not obliged to deal with problems of milk and cream transportation. His cream is always at hand and he knows just how much he will have to work with. These few advantages more than offset the superior equipment of the creamery. If the dairyman posts himself thoroughly on the principles and practice of farm butter making and puts out his product in attractive packages, he can compete favorably with the best creamery butter in the smaller markets and in the local retail trade. The average dairy butter is to-day of a much higher quality than ten years ago, while the converse is true of creamery butter. The general introduction and use of the hand separator has encouraged farmers to keep their cream a longer time on the farm than was formerly the case, and creamery butter makers are in despair over the low quality of raw material from which they are expected to make butter of merit. Creamery butter is living on its old reputation, and the recent drop in butter prices is due largely to a drop in quality.

A Few Essential Principles

The main factors in the production of fine butter are first, absolute cleanliness and, second, absolute control of temperatures. Milk to be used for butter-making purposes should be cared for with as much solicitude as when it is to be sold to a critical market trade. The utensils should contain no inaccessible places difficult to wash and sterilize, neither should they be rusty or used for any other purpose than in the dairy. Utensils which have contained milk should be washed first in luke-warm water to remove the milk solids, and then scalded and allowed to dry without being wiped.

If boiling-hot water is used on milk utensils before they are thoroughly cleansed the albuminous portion of the milk will be coagulated and made to adhere to the receptacle. Soap or washing compounds containing grease should not be tolerated in the dairy. Mineral cleansers only should be used. Sunlight is one of the best disinfectants known to science and exposure of dairy utensils to the bright sun daily will aid greatly in destroying the haunts of bacteria which produce bad flavors in butter.

Separation of Milk by Centrifugal Force

The first step in the manufacture of butter is the separation of the cream from the milk. For this purpose a centrifugal cream separator is the best and most satisfactory agent. The old method of allowing the cream to rise to the top of the milk by keeping it in a quiet, cool place has been superseded entirely by this more modern way. The method of setting the milk is also very expensive, since not over eighty per cent. of the fat is recovered from the milk as compared with about ninety-nine per cent. with a modern centrifugal separator. The separator furthermore clarifies the cream by removing mechanical impurities.

Still another advantage is the ability of the

operator to secure the richness or thickness of cream most desired.

Operation of the separator is simple, and the principle is but a development of the separation of the cream by gravity. The milk flows from a supply can into a rapidly revolving bowl. The speed varies from six thousand to twelve thousand revolutions per minute in different makes of machines. The cream being lighter than the milk is forced toward the center of the bowl, while the skim-milk occupies a position as far from the center as possible.

Separate spouts remove the cream and skim-milk, the process being almost continuous.

Care of the Cream

Milk separates best while still warm from the cow, although some machines will do fairly good work with cold milk. The best time for separating milk is immediately after milking is over, since the skim-milk will have a greater feeding value when fresh. As calves, hogs and other live stock make rapid and healthy gains on skim-milk this side-line of the dairy is worth developing.

When the cream has been separated, it is immediately cooled by placing the can containing it in cold water which is constantly being renewed. Warm and cold cream should never be mixed, as a disagreeable flavor is likely to be developed. All of the steps in handling the cream should be conducted in a sweet-smelling, clean and sanitary room or building. Milk rooms partly or entirely underground or near the barn-yard or manure pile will ruin the quality of the cream in a very short time.

The Process of Ripening

The purpose up to the present point has been to keep the cream sweet and clean flavored. If the cream should now be churned, a product known as sweet-cream butter would be secured. Owing to its rather flat taste, this kind of butter is not in demand with the average trade and is seldom found on the market in large quantities. The public desires a flavor in butter which is suggestive of mild "aging" or a "rich" flavor, as it is commonly called. In order to obtain this flavor, the cream must first be ripened. The ripening process is merely souring the cream in such a way that the lactic acid bacteria, previously mentioned, may develop. In order to secure the desired lactic acid flavor, a small amount of "starter" is now added to the cream. A starter is merely sour milk which has the desired flavor.

Making the Starter

To make a starter simply fill several quart glass jars with sweet milk, either whole or skimmed, and allow them to stand until the milk has soured. If the curd is solid and has a rather pleasant sharp taste, and if the curd becomes rich and creamy on being shaken up, the starter may be considered good.

If on the other hand it is full of gas holes and has an unpleasant, bitter or repulsive taste, undesirable bacteria have gained predominance over the lactic acid germs and the starter is worthless. A starter will develop best if kept in a warm room; it will generally be ready for use in about twenty-four hours. Starter may also be purchased in the form of bottled cultures, but for the farm dairyman, the home-made starter is recommended.

When the starter has become quite firm, it is ready for use. The amount which will be required depends chiefly on the condition of the cream. Cream that is very sour requires no starter at all if the cream is of a good flavor, but if the flavor is bad, a very large amount will be required to overcome the defect. The usual rule for adding starter is to use one-tenth as much starter as there is cream. A rather disturbing influence in the use of starter is the thickness of the cream. As it churns best when containing about thirty per cent. of butterfat, the amount of starter should never be so great that its addition will make a very thin cream. The ripening process requires constant study and a constant readjustment of rules and formulas if the best butter is to be made at all times.

Ripening of a batch of cream is most satisfactorily done in a large sanitary metal vat. Those used by creameries are provided with coils or compartments for cooling or heating purposes. Stirring is always beneficial, as it makes the cream ripen uniformly and maintains an even temperature throughout the entire batch.

Churning Temperatures

The best temperature at which to churn is about fifty-four degrees Fahrenheit in the summer time and sixty degrees in the winter. The cream should be kept at this temperature or slightly below it for about three hours before churning, so that the butterfat may have an opportunity to become as cold as the liquid portion of the cream. In practice, the butterfat is generally a few degrees warmer unless this precaution is taken.

The only step now to be considered before the operation of churning is the determination of the time that the cream is sufficiently ripe or "ready." Experience is the best teacher for this determination, but until the necessary amount of practice has been obtained, any of the widely advertised acid tests may be used. Cream is ready for churning when the flavor is sharply acid, though not bitter. This point will be reached when the acid test shows about six-tenths of one per cent. of acidity (.6).

Superiority of the Barrel Churn

Without question the best type of churn for use in the small dairy is the old-fashioned barrel churn. A large dairy where several hundred pounds of butter are churned daily may advantageously be equipped with a small size of factory churn. To prepare the churn for the cream first wash out with cold water, then fill about one-fifth full with boiling water and after clamping down the lid, revolve a few times. During this operation the draining plug should be out in order that the steam generated will not explode the churn. The purpose of this treatment is to thoroughly saturate the wood with water so that the butter will not stick to it. Now pour out the warm water and cool the churn by means of cold water or ice if it is available.

The Principle of Churning

Cream is now added until the churn is about one-third full. The exact amount of cream will depend somewhat on the shape of the churn, but never put more cream into the churn than can be thoroughly agitated. The concussion of the cream on the side and on the bottom of the churn is the force which welds the fat globules together and makes them into butter. If the churn is too full, the length of time required for butter to come may be more than twice as great as though one-half of the cream was churned at a time. The ordinary time for churning is forty-five minutes. If butter comes in less than this time, the temperature of the cream when added to the churn was too high, and the butter is likely to be of a weak body and greasy texture. If more than forty-five minutes is required for churning, the temperature of the cream was probably too low, and when the butter forms, it is likely to come in a hard, solid mass. Both extremes are objectionable and the operator should seek to avoid them.

Washing and Salting the Butter

When the butter globules become visible to the eye, it is a sign that churning is nearly completed, and when the individual lumps slightly exceed the size of wheat or rice kernels, the churning should be stopped. The buttermilk is allowed to drain off, but its quantity should be measured. Then an amount of water equal to the volume of the buttermilk is placed in the churn, the buttermilk plug being replaced in the meantime, and the churn should be revolved slowly. This does not result in any further churning of the butter, but washes out any buttermilk which may have adhered to the butter globules. This process is called washing the butter, and the temperature of the wash water should be about that of the buttermilk, at least no warmer, or some of the butter may become melted and washed away.

The amount of salt to be added will depend upon the demands of trade. One should carefully ascertain whether the purchaser desires a highly or lightly-salted butter. The average rule is one ounce of salt for every pound of butter. Add the salt dry but sprinkle well over the butter. Give the churn a few turns and allow to stand for several minutes. One of the most common faults in dairy butter is the uneven distribution of the salt and nothing can be done to remedy the evil after the butter has been packed unless it is re-churned, which is an expensive process and can easily be avoided by care in working the butter. The old method of working the salt well into the butter by means of a paddle has been superseded by table workers, or, in the larger churns, by rollers inside of the churn so that the butter can be worked without being removed from the churn.

The Process of Working

Working the butter causes the salt to be uniformly distributed and also affects the texture of the finished product. If a table worker is used, the butter is spread out on the bed and divided into two portions; one portion is placed upon the other and a roller is then passed back and forth until the butter has become smooth. An ideal texture is secured when the butter assumes a waxy appearance and feeling. Under no circumstances should working be continued until the butter has been reduced to a salve-like or greasy texture. Such a texture is known among commission men as a "weak body," and the best prices can never be secured for such a piece of butter.

The style of packing the butter for market must be determined by the kind of package most in demand. The old-fashioned stoneware crocks are suitable under some circumstances for private marketing or a local trade, but even such a package is not as attractive as a pound print neatly wrapped in parchment paper and then placed in either a cardboard carton or a wrapper printed in colors. Do not put any wrapping containing colors next to the butter, since consumers object to having printer's ink next to their food. Wooden boxes which will contain a considerable number of the prints are used for marketing butter put up in this form. Harmless artificial coloring matter may be used to give butter the shade of yellow demanded by the trade.

Establishing a High-Class Trade

For purposes of advertising as well as to insure customers against inferior grades of butter, the dairyman should have his name or trade-mark on every package he sells. Obviously only the very best butter should be put up under such a private brand, and in case a churning is inferior to the usual grade of butter produced, such a churning should be disposed of through some other channel.

The marketing of butter in rolls or ununiform packages will always result in lower prices and a less satisfactory demand. The dairyman who can always be relied upon for furnishing butter of a consistently high quality can charge more than the usual market price without injury to the volume of his business. The weight of any package of butter should correspond with the reputation of the dairyman. Packages which are short weight as little as half an ounce reflect on the honesty of the dairyman who will suffer in the end, either under the law or in the estimation of his fellow men. The weight of the wrapper or package, even though of light paper, should never be included in the weight of butter sold and paid for.

CHAPTER XVI

IS FARM CHEESE MAKING PRACTICAL?

In the development of the dairy industry of the world, the tendency has been to remove the manufacture of dairy products from the farm to specially-equipped factories managed by experts in their various lines. The more complicated the process, the more complete has been the removal. Taking condensed milk as one of the most difficult dairy products to make on a farm, we find that practically one hundred per cent. of the output of this product is made in factories. About two-thirds of the butter made in the United States is manufactured in creameries, and when we come to cheese, statistics show that of late years only about five per cent. of the total output in this country is made on farms, which indicates that cheese making is rather too difficult and complicated a process to be conducted to advantage on farms.

This conclusion is entirely justified, for with the demands of the regular trade for large quantities of cheese of uniform quality, age and package, the dairyman cannot hope to compete with the skilled cheese maker, who has many more conveniences to work with and who has made the manufacture of cheese a business. The preceding remarks apply to the making of those varieties of cheese which are found quoted in market reports and which go to our large cities in wholesale lots. Among these varieties are American or Cheddar cheese, brick, Swiss, Limburger and the manufactured soft cream cheeses. But outside of this list there are a few common varieties which are adapted to manufacture on the farm. In all there are some 282 different varieties of cheese, most of which are of European origin, and many are made from the milk of sheep and goats.

Varieties Adapted for Manufacture on the Farm

The varieties which can be made with the best assurance of success and profit on American farms are cottage cheese, sour cream curd (also called English cream cheese) and French cream cheese.

Cottage cheese is frequently referred to as Dutch cheese, pot cheese and schmiercase; all refer to identically the same product. Either whole or skimmed milk is used in its manufacture, but more commonly the latter. Set a large pan or vat containing the milk near a stove where the temperature will range between ninety and one hundred and forty degrees. In the course of a day the milk will curdle. The curdling process may be hastened by adding a small amount of sour milk. Allow to stand a few hours more until the curd has separated well from the whey and then drain through cheese-cloth. After the whey has drained off fairly well, the cloth strainer containing the curd may be squeezed gently to expel more of the whey. The curd is then washed with luke-warm water and next with cold water. This removes any particles of whey which may have remained and adds greatly to the keeping quality of the cheese.

Common Faults of Cottage Cheese

The curd is now lightly salted or seasoned with various spices according to the demands of the consumer. It may then be made into balls about two inches in diameter, packed in pasteboard buckets or put into other marketable form. Under no circumstances should cottage cheese be offered for sale as a semi-liquid mass. If the curd is not sufficiently dry to be handled easily and made into balls, the straining off of the whey was incomplete and the cheese will neither make a favorable appearance to the purchaser nor will it keep well. Properly-made cottage cheese can be sold in cities and villages for from ten to fifteen cents per pound. It is sometimes eaten fresh with the addition of a little cream or butter.

Sour cream curd, or English cream cheese, is made by straining the whey from thick, sour cream until the curd is dry. The straining is best accomplished by placing the cream in a strong linen bag which can be twisted to force out the whey. The dry mass is then salted and put into a convenient form for marketing. As cream is the basis for this kind of cheese, the profits derived are small unless the cheese is made only when sour cream is at hand and no other means of disposing of it are available.

French Cream Cheese

French cream cheese is the name applied to a wide variety of soft cheeses. The genuine product is made by first adding to fresh milk sufficient rennet to cause it to curdle in two to four hours. After being allowed to stand undisturbed for twenty-four hours, the whey is carefully poured off without disturbing the curd. The latter is then cut in slices and drained on a clean muslin sieve, after which fresh sweet cream is added and thoroughly mixed with the curd by stirring and mashing. The product may then be placed in small jars or wrapped in wax paper, which is in turn covered with tin-foil. This cheese should be eaten while fresh.

In marketing the varieties of cheese just described, the dairyman should study the nature of the demand and should endeavor to make his products of uniform excellence. Judicious advertising in local papers or by post card announcements to former customers will stimulate and create an interest in these products if the public does not respond at once to the opportunity of securing these varieties of cheese at a reasonable price. The consumer is generally eager, however, to vary his diet with something new and fresh and in many cases the manufacture of these simple varieties of cheese will be a profitable side-line of the dairy.

CHAPTER XVII

THE TRUTH ABOUT THE TUBERCULIN TEST

THE past ten years have witnessed the development and organization of civic bodies all over the country for the purpose of fighting and eradicating the disease tuberculosis, which is common to both man and many domestic animals. While the nature and the symptoms of the disease in the human and in the animal body are not identical, they are so similar that the question arose at an early date, "Is the disease of tuberculosis in animals, particularly the cow, transmissible to mankind?"

Eminent scientists disagreed for a time as to the correct answer to this question, as there was abundant scientific evidence to support both sides, but after several years of close observation by doctors and health officials, irrefutable evidence accumulated which showed that mankind, particularly infants and small children, can contract tuberculosis by consuming milk from animals which are affected with the disease.

Questions to be Considered

Tuberculosis is found in a small percentage of cattle all over the world, and wherever cattle are kept in large herds or indoors a great deal of the time the percentage is slightly greater than where the animals are on a large range. The importance of a thorough understanding of the disease is therefore evident if a dairyman hopes to be successful in keeping his animals healthy and expects to sell dairy products known to be free from death-dealing germs.

The main considerations in which the dairyman is interested are as follows:

What harm will tuberculosis do in my herd?

How can I tell if an animal has tuberculosis?

How can I get rid of it if I find it among my animals?

Is the tuberculin test reliable?

Why do some animals which have been condemned and killed show no signs of the disease?

Is there any danger of infecting a healthy herd by the use of the tuberculin test?

In case an animal or several animals fail to pass the test and are condemned, must the owner of the animal bear the total loss?

The first question "What harm will tuberculosis do in my herd?" can be answered thus: One animal affected with the disease can impart the disease to every other animal in the herd and probably to every human being who is working with the herd. Animals affected with tuberculosis will finally die of the disease, but not until they have lingered for a sufficiently long time to make their milk very dangerous. When the animals die, they are useless as food.

The second question, "How can I tell if an animal has tuberculosis?" is answered thus: The tuberculin test is the only definite means of establishing the presence of the disease in an animal.

Making the Tuberculin Test

The tuberculin test consists in injecting a sterile extract of tubercle bacilli underneath the skin of an animal suspected of having the disease. The extract cannot possibly contain any living tubercle germs and is therefore perfectly harmless. There is no effect upon the flow of the milk to any extent.

The presence of the disease is indicated by a rise in the temperature in the animal treated of three or four degrees. The greatest rise occurs about twelve hours after the tuberculin has been

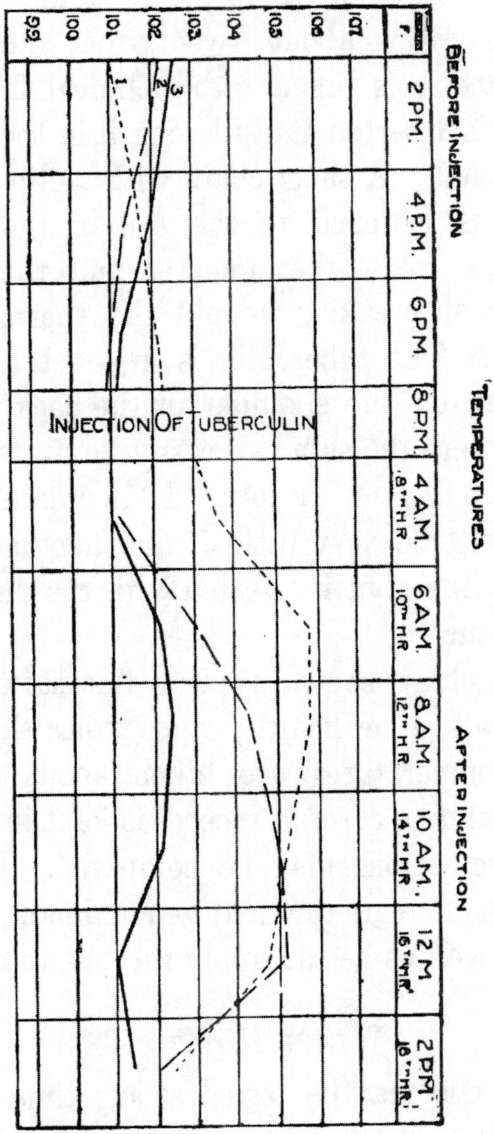

XVII.—Typical Temperature Curves; 1 and 2 are of tubercular cows and 3 of a healthy animal.

injected, and the temperature again returns to normal after about twenty-four hours. The temperature is taken with a clinical thermometer which is inserted its full length in the rectum of the animal. A short chain with a clip at one end which is fastened to the tail of the animal is attached to the thermometer and thus prevents danger of breaking should the thermometer be ejected. The tuberculin is injected just beneath the skin on the shoulder or the back part of the neck where the skin can easily be drawn together with the fingers in a fold. If an animal is healthy there will be very little if any fluctuation in the temperature, but if affected with the disease a rise will occur.

The chart shows typical fluctuations in the temperature of healthy and unhealthy animals. The temperatures are taken about every two hours and a careful record is kept in order that the fluctuations may be compared. In practice one person can test a dozen animals as quickly and as well as he can apply the test to only a few.

When Not to Test

Animals may be tested at any time except the following: when a cow is about to calve or has just dropped a calf; when an animal is suffering from an inflammatory disease or is otherwise likely to have a fever; when an animal is in heat. It is best not to apply the tuberculin test during extremely hot weather, since the temperatures are likely to be rather high from natural causes and may lead to the condemnation of an animal which is not affected with tuberculosis.

There is no way, other than the tuberculin test, of determining definitely whether a cow is affected with the disease. In the very advanced stages of the disease a cow will cough, show signs of unthriftiness, hold her head down and otherwise appear sick. These symptoms are noticed, however, after the disease has existed in her body for a great length of time. Yet even these are not sure indications that tuberculosis is the cause of her bad condition.

When the diseased animals have been determined, the sooner that they are disposed of, the quicker the spread of the disease in the herd will be checked. It takes several years of testing to be absolutely sure that the disease has been eradicated. The reason for this is the fact that at the time of the first testing, the tubercle germs may have just been received into the body of healthy animals but have not developed sufficiently to cause the animal to react to the test. When a second test, however, is made a year later, the disease is readily detected. As a rule a third test will fail to show diseased animals provided that those formerly condemned were quickly disposed of and no others were introduced without first passing the test.

Organs and Tissues Commonly Affected

The post-mortem examination of animals condemned will show the extent to which the disease has affected the body. The part most commonly showing lesions of the disease are the liver, lungs, mesenteries and connecting tissue, the udder and sometimes the windpipe. Any of the internal organs may be affected, also portions of the bone, muscles, and even the brain of the cow. The tuberculin test does not show the extent of the disease or its location but merely proves that the disease is present somewhere in the animal. If there is an uncertain reaction, a cow should be tested within three or four months so that she will not be in the herd a needless length of time. Do not run any chance of infecting the other animals with the disease.

Some of the many skeptical people who invariably stand in the way of progressive movements point to the fact that sometimes an animal is condemned for tuberculosis on the basis of the tuberculin test and on being slaughtered fails to show any indication whatever of the disease. Errors of such a kind are due not to any fault of the test, which is scientifically perfect and in itself accurate, but to errors in its application and manipulation. The operator who makes the tests should note whether the animals are excited and if so should ascertain whether their nervousness is sufficient to cause an elevated temperature. The operator should be competent to do the work and take the readings correctly. Unfortunately every one who tests cattle is not above the possibility of making an error, but mistakes of this nature should be charged up to human frailty and not to the tuberculin test.

Errors Not Due to Fault of Test

Furthermore the post-mortem examination of condemned animals is not always made carefully enough to detect the presence of the disease when it is present in the body outside of the places where tubercular growths are generally found.

This is also the error of the veterinarian and not of the test.

Laws of different states vary to a great extent in the disposal of the carcasses of animals affected with tuberculosis. If the disease has made but little headway, the carcass may be passed by duly-appointed inspectors and the healthy portions may be sold for beef. If the animal is badly diseased, it is generally buried in quicklime. In case the inspection and the testing have been done by a duly-appointed official representing the state, the dairyman, in some states, is reimbursed to the extent of three-fourths of the value of condemned animals. One should become familiar with the dairy laws of his state so that the greatest compensation possible may be received for tubercular animals disposed of under the regulations of live-stock boards.

How to Disinfect Dairy Premises

When the herd has been freed from tuberculosis in the manner just described, the barns and the premises should be thoroughly disinfected. Everything movable should be taken from the barn, and all smooth surfaces scrubbed, scraped and freed from dust and dirt. Then every square inch of surface should be thoroughly whitewashed with fresh lime. Burning sulphur is not an effective disinfectant. A saturated solution of corrosive sublimate (mercuric chloride) which can be secured at any drug store is also effective for spraying the walls and disinfecting manure heaps and litter which cannot be treated with whitewash.

Healthy Calves from Diseased Cows

If a condemned animal is a pure-bred or otherwise of great value, a careful breeder can raise healthy calves from such a cow, provided that certain fundamental precautions be taken to prevent the transmission of the disease from the mother to the calf. First, such a cow should be kept in absolute quarantine and no other animals should be allowed in the same building or pasture. The calves should be removed from the mother as soon as they have been dropped, and fed on the milk of other animals. Such management involves a great deal of work and continual care as well as risk to the healthy animals of the herd, though they are kept apart. In the great majority of cases, it is far better to kill every animal affected with tuberculosis.

The latter plan also gives purchasers of dairy

stock greater confidence when in search of healthy animals to add to their own herds. A dairy farm on which animals affected with tuberculosis are kept is always regarded with suspicion, and sales cannot be made to the best advantage.

CHAPTER XVIII

DAIRY ORGANIZATIONS

WHATEVER branch of dairying one may follow, he will be well repaid by affiliating himself with a number of the more important dairy organizations of the country. Milk shippers have their protective associations in order to secure fair prices for their milk and to be able to resist unfair treatment by middlemen and city distributors. Those dairymen who sell milk to creameries are often members of local creamery associations which sometimes direct the management of the creamery and very often own the building and its equipment. In any case, an association of the patrons of a creamery or a cheese factory has influence in securing for its members fair treatment and the best prices possible for their product.

There are also in many states dairymen's associations which look to the development of the dairy industry in general and to the protection of the industry against fraudulent substitutes, unjust laws and other abuses. Some dairymen's associations also are interested in the breeding of pure-bred stock and of assisting its members to detect and remove unprofitable animals from their herds.

Educational and Commercial Value of Dairy Meetings

If the dairyman is making a fancy grade of dairy butter, he should avail himself of the opportunity to have it scored or judged by experts in charge of the butter contests which are now held in many states in connection with dairymen's conventions and in dairy schools. County and state fairs frequently have displays of fine butter and the dairymen who exhibit their butter in such prominent places not only receive excellent advertising but may win valuable prizes usually offered.

Aside from the commercial value of associating oneself with other dairymen, a very important educational benefit arises from the discussions one hears at conventions and also from the personal contact with others. In addition to the benefits which may be derived from the organizations mentioned, a dairyman often requires expert advice on a wide number of subjects pertaining to his business. In a book of this nature a great many of the details of dairying must be sacrificed for the presentation of the greater and more important principles, but the following sources of information will be of invaluable help in solving some of the smaller perplexing problems.

Value Service Available from Federal Government

The Dairy Division of the United States Department of Agriculture is in a position to furnish persons engaged in dairying with general information of all kinds. Among the services which may be expected from the experts of the Dairy Division are personal letters in answer to inquiries, blue print plans for dairy buildings, advice as to the manufacture of dairy products, information concerning national dairy laws, bulletins dealing with practically all branches and phases of dairying and, if the importance of the request demands it, a special agent will sometimes be sent to the farm of the dairyman to render him personal assistance. In fact, nothing is so small that the Dairy Division will not give it prompt and careful attention. Inquiries may be addressed to either the Secretary of Agriculture, Washington,

D. C., or, if an immediate answer is desired, it may be sent to the Chief of the Dairy Division, Department of Agriculture, Washington, D. C.

Duties of State Dairy Departments

The Dairy and Food Departments of the different states are in a position to furnish dairymen with information concerning state dairy laws. Most of the dairy commissions are also compelled by law to investigate and act on violations of the law reported by reputable persons. Such a requirement makes it possible for a dairyman who is in competition with a person who is using fraudulent methods or who is selling his products to a creamery which he believes is defrauding him in payment, to have the matter investigated. In this manner those who are conducting a legal business are protected and those who are violating laws are either put out of the business or are compelled to improve their policy. The dairy commissions generally have inspectors who can be called upon to visit one's farm and make suggestions for improvements from a sanitary point of view.

Advantages of Dairy School Assistance

In nearly every state of the Union there is a dairy school the members of whose faculty are well qualified as experts and as educators. Correspondence addressed to the professor of dairying at such a school will receive about the same attention given by the Dairy Division already described. The particular advantage of dairy school assistance lies in the fact that the latter is nearer, is more quickly obtained and that the professors of such a state institution are for the most part more familiar with local conditions than are the national authorities whose scope is much broader and therefore their services must be along more general lines.

In case one is interested in extending his knowledge of the science of dairying particularly, correspondence may be addressed to these dairy schools, which will supply information regarding the courses of instruction given. During the winter months there are generally a number of short courses ranging in length from one to fourteen weeks and covering branches extending from dairy herd management to the operation of machinery and tests for dairy products. As most of these courses are given about the middle of winter when other work on the farm is not pressing, a great many progressive dairymen avail themselves of the opportunities afforded.

Services Extended by Dairy Publications

The editors of dairy papers, of which there is a considerable number in the United States, are in a position to furnish information concerning the best makes of churns, stanchions, manure carriers and the great list of sundry equipment, which is needed on every dairy farm. In order to get the quickest and most satisfactory service, one should be a subscriber to the paper. The advertising matter of first-class dairy papers is always reliable, and information of the kind mentioned may be obtained directly by writing to advertisers and securing their catalogues. Breeding animals may also be obtained by looking over the cards of breeders in dairy-paper advertising matter, and corresponding with those breeders who seem to have the best selection. Books on some of the detailed subjects connected with the dairy industry, such as ice-cream making, and manufacture of fancy cheese, may also be obtained through the dairy papers. The editors will also take care of inquiries on questions which may be asked of them.

Information on Animal Diseases

In a great many of the dairy states there is a sanitary board which deals with animal diseases and the suppression of epidemics. In case of the breaking out of a disease in the dairy herd or if an animal requires prompt medical attention, correspondence should be addressed to this board, which will either furnish a veterinarian or will give the address of a competent veterinarian in the nearest town. The post-office address of the sanitary board is generally at the state capital and can always be secured by writing either to the state dairy school or to the editor of a state dairy or agricultural paper.

Interest is centered once a year in the National Dairy Show, which generally is held some time during the fall in one of the principal cities of the central West. The show extends over a period of about ten days and during this time most of the prominent national dairy organizations hold their conventions. In connection with the Dairy Show there are exhibits of cattle, dairy products, machinery for dairy purposes, supplies, demonstrations and addresses by national dairy authorities on questions of vital interest. The show is attended by progressive dairymen from all parts of the country and the benefits derived from attending will far outweigh the expense.

The exact date and the place for holding the Dairy Show is announced by all of the prominent dairy papers about the latter part of July. A list of the meetings and of the organizations which will convene at that time may be obtained either from the editors of dairy publications or from the Dairy Division of the United States Department of Agriculture.

CHAPTER XIX

HOW AND WHY THE DAIRY MARKETS FLUCTUATE

In whatever branch of dairying one may be engaged, his income will be affected by the condition of the markets for dairy products. Of all the principal products, milk fluctuates the least in value when it is sold for direct consumption. Raw milk retails in the cities of the United States for from five to ten cents a quart. It is cheapest in the central West and dearest in the extreme East and on the Pacific Coast. The average for the entire country is about seven cents a quart. During the winter months, the price is from one to two cents a quart higher than in the summer. In the far South the purity of the milk supply is dependent on the great problem of keeping it at low temperatures, and the expense of keeping the milk cool of course adds to its ultimate cost to the consumer.

The reason why milk is higher during the winter than in the summer is the higher cost of feed and the increased quantity required to maintain a liberal flow of milk. The higher cost of milk in the East than in the central West is due first to the more rigid sanitary requirements to which the dairymen must conform and second to the greater distance which milk must be shipped. The prices mentioned are those the consumer pays. The net returns to the dairymen are more nearly the same all over the country. Cream prices for the city trade vary about the same as milk, cream being usually four times as expensive as milk, volume for volume. Unless the dairymen supplying milk to a city or town have some form of organization, they are powerless to determine what prices their product is to bring.

Market Milk Prices are Governed by Local Factors

The city milk dealers usually fix the price, or if the dairyman delivers his milk from his own wagons, he will be boycotted by the public if he advances his prices over those charged by other distributors. The public has not yet learned to discriminate between good and bad milk, and will generally purchase the lowest-priced milk, provided it does not seem objectionable in any way. The latter remarks apply chiefly to the smaller towns where milk inspection is not actively carried on. Obviously the market for milk and cream for direct consumption depends very largely on the attitude of the public toward the milk supply and upon local conditions of supply and demand. These must of course be considered before engaging in the business near any particular town for the purpose of contributing to its milk supply. As a general proposition, the value and the prices of milk for direct consumption are increasing everywhere in the country.

The Three Great Butter Markets

The price of butter and of milk and cream sold to creameries depends on three large wholesale markets, New York City, Chicago and Elgin. The commission merchants of these markets have their finger upon the pulse of the public. They know about how much of each grade of butter their trade will require each month of the year. They know how much butter is in cold storage in all parts of the country and they are good guessers, for the most part, of the price the public will pay. The New York market is about one cent higher on the average than the Chicago and Elgin markets owing to the greater distance which butter must be shipped. The bulk of all creamery butter is made in the central West.

The Situation in the West

On the Pacific Coast, the prices are somewhat higher than in the East, but the geographical lay of the land and the difficulty of shipping butter to the coast cities have contributed to the high prices which do not promise to be permanent owing to the rapid development of creameries and dairy herds in recent years. The large ranches are fast being broken up into smaller holdings on which dairying is an important branch of the farming being conducted.

As milk or cream sold to creameries is paid for, with very few exceptions, according to the amount of butterfat it contains, and since the market value of butterfat varies directly with the price of butter, the interest of the producer in the market situation is evident. The only way in which the producer can influence market prices is by selling to the creamery only the highest grade of product, thus enabling the creamery to make a uniformly good grade of butter. There is *always* an active demand for good butter at satisfactory prices all around.

Cheese Factories as Markets

Milk sold to cheese factories is paid for in a number of ways, the method being generally determined by the patrons. Perhaps the most common method is by weight, on the basis of one hundred pounds. For example, if the current price is $1.40 per hundred, each patron receives this rate of payment regardless of the richness or purity of his milk. The only reasonable explanation for the justness of this method of payment is that normal milk does not vary sufficiently in its value for cheese making to demand an analysis of its constituents, on the basis of which payment should be made. Thus the rather approximate method of paying for cheese milk by weight continues to be practiced.

Another method of paying for milk delivered to cheese factories is according to the Babcock test in the same manner as is done by creameries. The milk is weighed and tested for its butterfat content. The weight of the milk is multiplied by the test which gives the number of pounds of fat and the product is multiplied by the current price for butterfat. The prices paid by cheese factories are determined by the prices secured for the finished product. The latter are determined by the condition of both the market and the cheese. Cheese markets are, as a rule, more stable than butter markets, though fluctuations may occur when least suspected. The chief objection to paying for cheese milk according to the Babcock test is that the value of milk for cheese making depends not upon its content of fat but upon both the fat and the casein. In addition to a difference in flavor caused by bacterial action, cheese differs from butter in that the former consists largely of casein or albuminous matter, whereas butter contains but a fraction of one per cent. of such a constituent. Therefore the most rational manner of paying for cheese milk is according to both the fat and the casein tests, or as they are more commonly called, the Babcock and the Hart tests, each bearing the name of its originator.

The Danger of the Condensery

Another market for milk which varies considerably is the condensery, or plant in which milk is manufactured in the various kinds of condensed, evaporated and prepared milks put up in tin cans. Most of the condenseries in the United States are located in New York, Pennsylvania, northern Illinois, Wisconsin, Washington and Oregon. The condenseries are for the most part controlled by a few persons or companies, and the farmer has no means of determining the prices he is to receive except by organizing to demand higher prices when such demands are justified. Milk condenseries pay slightly higher prices for milk than do either cheese factories or creameries, but as the producer receives no by-product such as skim-milk, buttermilk or whey, he is poorer in the end than if he received slightly less for his milk and retained such valuable feeds as those mentioned for feeding purposes. Experience has shown that milk condenseries finally result in run-down farms and herds. In fact the sale of the whole milk from the farm is almost as ruinous to its fertility as the sale of wheat and grains rich in valuable mineral elements. The maintenance of fertility on the dairy farm will be discussed in another chapter.

CHAPTER XX

WHY OLEOMARGARINE IS INFERIOR TO BUTTER

COMPETITION makes and unmakes businesses. The competitor of any business man either forces him to put out better goods or to sell on a closer margin if he expects to hold his trade. A new invention may ruin many business enterprises. The telephone has made serious inroads on the messenger service business and the phonograph has practically put an end to a demand for the old style music boxes. So in the dairy business, one may ask, "What are the chances for butter substitutes ruining the butter markets and lowering the prices to dairymen?"

The answer to this question is simple and final. Oleomargarine and the various similar products can never replace butter on the market nor will they interfere seriously with the demand for butter because they are inferior products. True, at the present time, our oleomargarine laws are lax and inefficient, but the public is now demanding more stringent regulations, which will prevent the fraudulent sale of oleomargarine for butter. The only way in which the manufacturers and distributors of oleo have been able to advance the sales of their products is to sell them for butter or in such a manner as to make the purchaser believe he is receiving butter. Oleomargarine is offered for sale under such misleading terms as "fancy creamery, Jersey brand" and similar names which do not include the term "butter," yet convey that meaning to the consumer. In many restaurants, oleomargarine is also served instead of butter without warning to the diners that they are receiving a substitute.

Relative Value of Butterfat and Oleo Fats

Oleomargarine is the official and general name for all butter substitutes made of animal fats. Butterine, margarine and oleomargarine are the same commercial product. They are made by combining in various ways animal fats with various oils of which cottonseed oil is a typical one, and agitating the mixture in milk, cream or sometimes soft butter in order to permit the fatty mass to take on a butter flavor. Oleomargarine is made of cheap substances and is therefore a cheap product. The fats and oils of which oleomargarine is composed would be difficult for the human stomach to digest in their original form and, even in their finished form, their digestibility is far inferior to that of butter. The fat contained in butter is more nourishing, more easily digested and more palatable than the fats of which oleomargarine is made. Therefore, butter is worth more and is a better product than its substitutes, just as gold is superior to the various yellow metals sold at a low price because they are of less value than gold.

If equal quantities of butter and oleomargarine are placed in an oven with a glass front and heat is gradually applied, the butter will be observed to melt much more quickly than the oleomargarine. The melting point of butter is below blood heat. For this reason butter becomes a liquid in the stomach and its valuable constituents are easily extracted by the digestive juices. On the other hand, oleomargarine melts at a temperature several degrees higher than that supplied by the human body and is a solid in the stomach. Its digestibility is therefore obvious and, in European countries, where careful study has been made of its digestibility, it is barred from use in a great many hospitals and institutions for persons of feeble health even though the latter are provided with their living at the expense of the government.

The Intent of Oleomargarine Laws

Under the present United States laws (1911) oleomargarine cannot be colored and sold unless a tax of ten cents per pound has been paid to the Bureau of Internal Revenue. Uncolored oleomargarine is taxed one-fourth of a cent per pound. These taxes are not designed to add to the cost of the product and to favor the dairy industry by so doing, but rather to protect the public against fraud. The ten cent tax was designed to keep the yellow color, which is a natural trade-mark of butter out of butter substitutes, thus enabling the public to readily distinguish between the two. The use of yellow coloring matter in oleomargarine is one of the greatest compliments which can be paid to butter, as it indicates that, even by the payment of the ten cent revenue tax, the manufacturer can make a profit from his inferior goods, since it is sold on the reputation of butter. The tax of one-fourth of a cent per pound on uncolored oleomargarine is simply a means for compelling the manufacturers of this product to bear the expense of the inspection which is necessary in order that the public may not be defrauded.

Butter Market not Seriously Threatened by Oleo

The above taxes are, of course, subject to constant changes as the means of controlling the butter substitute industry by legislation become better known. The public does not lean toward the purchase of oleomargarine as oleomargarine. Although some, particularly the poorer classes, know what they are buying when they buy oleomargarine, the better class of consumers may always be expected to buy butter. When our oleomargarine laws have been made sufficiently strong to hold that substitute on its own ground, oleomargarine will not affect the butter market to any greater degree than will such products as apple butter, peanut butter, preserves and jellies.

CHAPTER XXI

THE CREAMERY PROMOTER AND HIS TACTICS

OUTSIDE of stock feeds, the dairyman is not often imposed upon by commercial quacks or separated from his money by dairy get-rich-quick schemes. There is one subject, however, which every dairyman should understand, namely, the manner in which the creamery promoter operates. In order to be a financial success, every creamery should have available the milk of at least four hundred cows, and should be so managed that the butter can be made for not over four cents per pound as the entire cost of manufacture. This cost includes depreciation of the building and equipment, salary of butter maker, cost of salt, packages and supplies. In very large creameries, butter can be manufactured for from one to two cents per pound. As these figures are much less than the dairyman can afford to make up his own butter for, the creamery will be well supported and will pay the farmer very satisfactory prices. But how about the section with no creamery?

Along comes the professional creamery promoter into a district where quite a number of farmers keep cows. He describes to the dairymen the prosperity and the financial success of creameries in adjoining counties and offers to assist the dairymen in establishing a creamery of their own as a market for their milk or cream. Many of the local farmers have doubtless heard of the success of the creameries mentioned by the promoter and, believing that other statements which he makes are also true, they fall in with the promoter's plans.

The Prices are Inflated

He tells the farmers how much the plant and its equipment will cost, and furnishes them with information on running and managing the creamery. The farmers who are unfamiliar with the current prices of creamery equipment and the actual cost of the necessary building are induced to sign notes or otherwise to obligate themselves financially.

The creamery promoter, by the way, is invariably a representative of some building- and supply-company, and government statistics show that the prices charged the farmers are nearly double the sum for which the farmers could have, themselves, built and equipped the plant. In addition to charging exorbitant prices, the promoter induces the farmers to put in a great deal of unnecessary machinery. The machinery absolutely necessary is also very frequently of a larger size than the butter-making plant will require for years to come. Frequently an attempt is made to find out the number of cows in the community from which milk may be secured for butter-making purposes. A very common error in making this cow census is to count all of the cows in the herds, including those not giving milk and also a great many beef animals, which give milk but a very short time during the year.

Why Promoted Creameries Generally Fail

Now for the outcome. The promoter departs with notes in his pocket aggregating upward of four thousand dollars and leaves the butter-making plant in the hands of the dairymen. Under the circumstances the farmers will do exceptionally well if they are able to make such a plant financially successful. A few such creameries have, after several years of skilful management, succeeded, but the majority have failed. The chief reason, as before suggested, has been the small volume of milk received and the consequent expensive cost of manufacturing the butter.

The second reason has been a lack of organization among the patrons of the creamery, and the third, a misunderstanding of the best manner in which to put their product on the market. After a few months, the farmers begin to realize that the promises of the promoter for high prices for their milk and big dividends by the creamery were without foundation, and the dissatisfaction which soon follows generally results in the dissolution of the creamery organization and the sale of the building and equipment to the highest bidder whose offer is very low compared with the amount represented by the farmers' notes to the promoter. Not only are the chances for a really prosperous and well-managed creamery ruined in that locality by the tactics of the creamery promoter, but a damper is placed on dairying and the creamery business for miles around.

Organizing on the Coöperative Plan

By far the best and always the safest way in which to organize a creamery company for the building of a butter-making plant is to begin on a coöperative plan. The product of four hundred or more cows should be insured for manufacture into butter, and marketing facilities should be carefully investigated. Information and blueprint plans for the cost and construction of the building should be secured from the Dairy Division at Washington, D. C., or from some other equally reliable source.

An expert butter maker will have to be employed in order to compete with other creameries for quality and market trade. Such a buttermaker will demand a salary of eighty dollars or more per month, but a good butter maker at one hundred dollars is cheaper than a poor one at sixty dollars. The losses of butterfat in the buttermilk through unskilful handling or the low percentage of overrun obtained through misunderstanding of the technical points may mount to the sum of thousands of dollars in a year. Even though an expert butter maker can be obtained at a satisfactory wage scale, the dairymen who supply the creamery with milk or cream will do well to post themselves on the business management of the plant, especially in regard to the bookkeeping and marketing end.

Customary Method of Marketing Butter

Creameries manufacturing two thousand or more pounds of butter weekly find it most profitable to ship their product to reliable commission merchants or wholesale buyers in the large market centers. The ability of such dealers to sell the butter to the best possible advantage more than offsets the charges made for the service.

CHAPTER XXII

DAIRY FARM FERTILITY

In every branch of farming one must consider both the visible and the invisible profits. The visible profits are easily determined by the status of one's bank account and the proceeds from the sale of products of the farm. The invisible profits are those arising from an increase of fertility, from the increased value of the farm due to better roads or traction facilities or from the building up of a progressive town in the vicinity. Visible profits are sometimes the only ones considered, but for an accurate determination of the real success of the farm, both the invisible and the visible profits or loss must be determined.

Frequently the visible profits offset the invisible losses and make the farm appear profitable when, in reality, such a conclusion is unjustified. A run-down wheat farm may apparently be giving the owner a reasonable income, whereas the depreciation in the value of the land is greater than the sales of the wheat. A good farmer will always recognize such a trend of conditions before they have advanced too far and will take steps to correct the points in which his farm is deficient.

Put Back What You Take Away

The state of Wisconsin was a run-down wheat state twenty years ago, and New York soil was also run down from cropping heavily with various farm and truck crops. Both states have gone into dairying extensively yet, at the present time, Wisconsin's soil is a great deal more valuable for agricultural purposes than that of the Empire State. The reason for the difference in these two prominent parallel cases is a very interesting chapter in dairy farm fertility. The writer realizes that general conclusions cannot be drawn which will cover every square mile of agricultural land in each state, but the chief cause for a lack of fertility in New York and for the restoration of fertility in Wisconsin has been the kind of dairying most common in each of the two states. Wisconsin is a great butter and cheese state; New York produces in addition to some butter and cheese a tremendous amount of milk for the large eastern cities and the milk condenseries.

Every ton of butter made takes but fifty cents' worth of fertility from the farm on which the necessary raw material for the butter was produced. If, on the other hand, the amount of milk required to make the butter had been sold to the city trade, about eight dollars' worth of farm fertility would have gone with the milk. Now, while the sale of milk from a farm does no injury to the land provided an equal amount of fertilizing matter is restored, the failure of each farmer to recognize this important point and to put back the fertility causes and has caused large sections of country to depreciate in value.

Barn-Yard Manure Better Than Most Chemical Fertilizers

The by-products which the dairyman receives from a creamery or cheese factory in the form of buttermilk, skim-milk or whey enables him to raise more live stock which, in turn, produce more manure. As a ton of manure is worth two dollars for fertilizing purposes, a large production of it on every farm indicates good management and large invisible profits. The majority of careful experiments made by agricultural scientists in testing the relative value of various kinds of fertilizers show that barn-yard manure stands very close to the top of the list. A few expensive and concentrated commercial fertilizers result in greater immediate returns, the first year or two after they are applied, but the noticeable value of barn-yard manure extends over a period of nearly ten years.

In addition to its fertilizing properties, it has a beneficial effect on the texture of the soil, when the manure contains straw or litter. In order to produce the best results, manure should not be allowed to remain uncovered in the barn-yard. If such a practice is followed, the rains will leach out a large percentage of the soluble mineral fertilizing constituents and the action of the sun and winds upon the manure pile will cause it to undergo fermentation and decomposition. The latter results in the loss of ammonia and other gases rich in nitrogen, a valuable fertilizing element.

Application of Manure to Fields

Manure is best handled by being hauled directly to the fields after it is produced. When such a practice is not convenient, the manure should be stored on a cement or liquid-tight floor and should be sheltered with a roof which will adequately protect it against sun and rain. In order to prevent the loss of volatile gases, such as ammonia, a small amount of ground phosphate-rock should be added to the manure pile every day. This will fix most of the gases produced in a firm chemical union and retain them for future use by crops.

In case skim-milk is fed to farm live stock, the dairyman need have no apprehensions that he is

XVIII.—A PROMISING BUNCH OF CALVES raised on skim-milk from the separator.

causing his farm to depreciate in value from a loss of fertility. Buttermilk and whey have about one-half the combined feeding and fertilizing values of skim-milk and when these are fed, soil fertility will need attention though there is little danger of the soil becoming worn out in case manure is conscientiously used. But if milk is sold to condenseries or to the city trade and no by-product whatever is retained on the farm, radical steps must always be taken to restore the fertility of the dairy farm before it is so run down that normal crops cannot be raised.

Two Theories of Soil Fertility

The two theories of soil fertility are, first, that every acre of land contains a certain amount of mineral plant nutrients and if the amount of these is reduced, the crops will suffer. The second theory is that the earth is a storehouse of mineral wealth which is constantly being distributed by the ground waters and that no loss of fertility will ever result where there is abundant moisture in the soil, but every crop leaves in the soil a mild toxin or poison which will reduce the yield of another crop of the same kind. For this reason no one crop can be grown on the same piece of land year after year with equally good results but, if a rotation of crops is followed so that a crop of corn follows a crop of clover and oats follows the corn, all of the yields will be good, since the toxins left in the soil do not affect a crop of a different kind.

Practical and experimental evidence indicates that both theories are well founded, and with an understanding of each, the dairyman who values his business or his farm will not allow it to depreciate in value.

CHAPTER XXIII

DAIRY ACCOUNTS

IF the dairy farm is well chosen with respect to markets and to its ability to grow the crops necessary for feeding, the dairyman should not only make a comfortable living but should receive good returns on his investment. If the profits of the farm are low and unsatisfactory, its owner or manager should be in a position to definitely locate the trouble and apply a remedy. A careful record of dairy accounts is therefore of the utmost importance for success in dairying.

No detailed outline for dairy farm bookkeeping will be suitable for every need, but the following points should be included in the computations, whatever system may be adopted. At the beginning of each year a careful inventory should be made. This inventory should include a live stock census, young and old animals being listed separately, a record of machinery and supplies on hand, and of course the value of feeding material harvested from the fields. In addition to the above figures, which can be made very definite, a record should be kept of the number of acres in the farm devoted each year to the different crops, pasture and unused land, if any. The inventory should also include the value of houses, barns, sheds, fences and bridges. The above listed items indicate what the dairyman has to work with, and they show his investment.

An Outline of Milk Records

In order to accurately determine the income from the herd, one should procure a well-bound book, such as a ledger, for keeping herd records. Every animal in the herd should be given a name, number or be so designated that any profit or loss may be definitely traced to her. At every milking her milk should be weighed and the weight recorded in the book which has been so ruled that a monthly record can easily be kept of her volume of production. Once a month the milk should be tested for its percentage of butterfat, and the total production of this determined by multiplying the amount of milk by its test. The most accurate method of testing the milk is to take a composite sample from ten or more milkings at different times during the month. The samples are placed in a tightly stoppered bottle which also contains a preservative to prevent the milk from curdling. Such preservatives can be obtained in tablet form from any dairy supply house.

GETTING AT AN ANIMAL'S NET VALUE

Having found the total production of milk and butterfat of each cow for a month, the dairyman should now compare the results with the amount and the cost of feed consumed by each cow. Pasturage, which is of course included in the computation, is commonly valued at fifty cents a month for each animal. The other feeds should be valued according to their cost, if purchased, and according to their market value if raised on the farm. The study of production- and feed-records will guide the dairyman in the selection of the most suitable feeds and in changing the ration of such cows as appear to be doing poorly as economical producers.

Obviously at the end of the year, the relative merits of every animal will be literally an open book and unprofitable cows may be disposed of or fattened for beef.

Together with the milk and butterfat record, a

breeding record should also be kept. In the latter the dates of breeding and calving are recorded and the value of the calf is credited to its mother. By adding the commercial value of the calf to the value of the milk, the gross income from each cow is determined. In this calculation the value of manure is overlooked, since manure is more properly included in the general farm inventory. Yet to be very exact, one may multiply the number of loads of manure, hauled to the fields, by two dollars and divide the product by the number of animals which produced the manure. The quotient represents the value of the manure from each animal.

Bookkeeping Removes Dairying from the Class of Manual Labor

In a separate part of the ledger an account should be kept of wages paid hired help, both regular and casual. Other expenses for labor, such as amount paid to haulers of milk and cream, services of veterinarian or similar expenses are most conveniently kept in this department of the accounts.

Still another portion of the account book should be devoted to the gross income, including returns from products and by-products sold, sales of live stock, feeds or compensation for services rendered to others.

Another list of importance includes general expenses, such as repairs, improvements and miscellaneous purchases. Now by uniting all of the credit and the debit items, the net income is readily determined and by noting its relation to the last annual inventory the dividend on the investment is ascertained. Without an accurate system of accounting, the satisfaction of conducting a dairy or any branch of farming quickly disappears and farming is reduced to mere manual work accompanied by the hope that everything will turn out all right financially.

Locating a Profitable Side-line

Bookkeeping shows whether the work is being conducted along the best lines. Quite frequently a side-line of the dairy proves nearly as remunerative as the main branch, thus indicating that the side-line can be profitably developed and perhaps made to be the main source of income. Owing to its complexity, farming offers excellent opportunities for wise and careful managers who are willing to study the business end. By mastering details and stopping leaks and petty losses, good managers have made small farms more profitable than vast ranches carelessly managed.

CHAPTER XXIV

DAIRYING FOR HEALTH AND WEALTH

About the first step toward the improvement of a business man's health after he has impaired his vigor seeking a financial fortune in the city is to establish a country home and to engage in some form of dairying, generally the production of a high grade of milk. These men of business habits see in dairying not only an excellent opportunity for supplying pure milk, so sadly needed in large cities, but a profitable business enterprise as well. Few sights are so beautiful and typically rural as a herd of cows on pasture, and few enterprises combine opportunities for health and wealth as satisfactorily as dairying.

In certain parts of the East, milk farms have been established as health resorts and the persons sent to these farms are required by their physicians to drink large quantities of milk and buttermilk. Few beverages contain the mild acidity, so beneficial to the human stomach, that is found in buttermilk. That age brings a quick deterioration of market milk has long been established and, for this reason, persons who expect to derive the greatest benefits from a milk diet should consume the milk as quickly as it can be cooled when fresh from the cow. In many other respects the quiet atmosphere of the dairy barn with its soft-eyed occupants is one of the best restoratives for a broken-down nervous system. These may be some of the reasons why dairying is so popular with persons who have succeeded in business life in cities and seek agriculture as a pleasurable and commercial supplement to their regular business.

The production of milk on farms is not a competitive business and there is room for all. Mother Nature is a silent partner in the business and by studying her methods and laws, your partnership with her will be successful.

CHAPTER XXV

DAIRYGRAMS, PROVERBS OF UNLIMITED APPLICATION

Honesty begets confidence and confidence begets business.

The dairyman who maintains sanitary premises at all times need not dread the visit of the inspector.

Ignorance of the law excuses no dairyman. The highest development of the dairy industry will be accompanied by the highest development of the law.

Gentleness shown the dairy cow pays large dividends in the form of liberal milk production, more valuable and more salable animals and more congenial work in the dairy.

However good are his intentions, a dairyman is judged by the quality of his products. Dairy products are never any better than the material from which they are made.

A dairyman who suspects the presence of tuberculosis in his herd and refrains from having his animals tested for fear of possible financial loss is as dangerous to civilization as a criminal.

Organization is the best tool by which dairymen may fight against discriminatory competition; education is the best means for overcoming ignorance and unfounded prejudice.

As long as the human stomach craves for food, milk and its products will be in constant demand. Owing to its perishable nature, the milk market can never be cornered by a few. Dairying will always be a favorite line of farming for men of limited means.

The corner-stone in the foundation on which pure milk supplies are established is cleanliness. Cleanliness in dairying demands the elimination of invisible contamination as well as visible dirt. Success in dairying means success in a struggle against contamination.

CHAPTER XXVI

SOME COMMON DAIRY TERMS

THROUGHOUT dairy publications and literature of technical nature, various terms and words are used which are found neither in ordinary literature nor in a great many of the best dictionaries, in the sense in which the terms are used in the dairy industry. In order to give the reader of a non-technical book of this nature an insight into the meaning of terms which he will be likely to encounter, the following list has been prepared:

TERMS APPLIED TO CATTLE

Pure-bred.—An animal whose parents were both typical representatives of a well-recognized breed; full-blooded. The term "thorough-bred" is applied chiefly to horses and dogs, never to cattle.

Grade.—An animal, one of whose parents, usually the male, was a pure-bred and the other of no well-defined breed. The term "grade" is often used as an adjective preceding the name of a breed, for example "grade Jersey." If a grade-

Jersey cow is mated with a pure-bred Jersey bull, the calf will be of a higher grade than the cow and if the breeding process is continued several generations, a high grade Jersey herd may be developed. When the calves are fifteen-sixteenths full-blooded, the next generation may be called pure-bred.

Cross-bred.—The result of mating a pure-bred animal of one breed with a pure-bred animal of another breed.

Scrub or nondescript.—An animal neither of whose parents was a pure-bred.

Terms Applied to Butter

Adulterated.—Butter which contains sixteen or more per cent. of moisture ; butter which contains any preservative other than common salt ; butter which does not conform with the legal requirements of a particular state. It is obvious that butter may be classed as adulterated in one part of the country and be legal in another section under a different code of laws.

Country.—The term applied to butter made on farms in contrast to that made in creameries.

Creamery.—A term applied to butter made

from the product of more than one herd and generally in an especially equipped factory, the machinery of which is operated by power as contrasted with machinery operated by hand.

Dairy.—The term applied to butter made in a farm dairy, dairy butter being generally superior to country butter, since a dairy is understood to be a place where adequate means and suitable machinery are at hand for the manufacture of butter. Dairy butter is also defined as butter made from the product of a single herd of cows.

Extra.—A market term signifying the highest grade of butter in the West and central West and the second highest grade in the East. Special is the term used for the highest grade of butter in the East.

Greasy.—The condition of butter caused by churning at too high a temperature or by overworking; a soft and salvy condition.

Mottled.—An ununiform appearance of butter caused by the uneven distribution of salt which makes some parts of the butter appear lighter in color than other parts.

Print.—An oblong block of butter weighing a definite amount, generally a pound, though sometimes two pounds. The print derives its name from the fact that the name of the manufacturer or distributor is either impressed in the butter or is printed on the wrapper.

Renovated.—Butter which has been subjected to melting and treatment with a blast of air and other purifying means calculated to remove rancidity and objectionable odors. Renovated butter is generally made from a poor grade of country butter.

Square.—A term applied on the Pacific Coast to a package of butter weighing about two pounds.

Sweet-cream.—Butter churned from cream which has not been allowed to ripen or become sour. Such butter has a rather flat flavor, but is in demand by certain classes of trade.

Tub.—A term applied to butter which is sent to the market in wooden tubs, generally holding about sixty pounds.

Uncolored.—Butter to which no artificial coloring matter has been added. Uncolored butter, while rather yellow during the months of May and June, is almost white during the winter months when cream is very slightly colored by the natural color of the butterfat.

Unsalted.—Butter to which no salt has been added. Applied in a more general way, unsalted butter is butter which contains a very slight percentage of salt.

Terms Applied to Milk

Adulterated.—A term applied to milk which is below the legal standard of milk solids or fat or which has been illegally skimmed, preserved or treated.

Blue Milk.—A term used to describe the bluish appearance of skim-milk. The bluish tint is produced by the action of a color-forming species of bacteria.

Buttermilk.—The product which is formed when butter is churned. Buttermilk, when fresh, is mildly acid, is of a yellowish-white color and is a favorite hot weather beverage.

Certified.—Milk whose purity and the sanitation accompanying its production has been officially certified to by a reputable board of citizens, doctors or milk experts.

Cheese-milk.—Milk which is intended for manufacture into cheese.

Clarified.—Milk which has been passed through a centrifugal separator or otherwise been subjected to sufficient centrifugal force to remove impurities suspended in the milk.

Condensed.—A form of milk produced by the removal of about two-thirds of the water constituting normal milk. Condensed milk is usually sterile and is placed on the market in tin cans. It may or may not contain sugar.

Dipped.—Milk which is delivered to the consumer by being dipped out of cans or large receptacles and transferred to the consumer's vessel. "Dipped milk" is a term commonly used in contrast to bottled milk.

Dried.—Milk which has been reduced to a dry powder by the evaporation of the moisture and drying.

Homogenized.—Milk which has been subjected to a treatment which breaks up and emulsifies the fat globules so that they will not rise. This treatment is intended to make thin milk appear rich but has certain commercial merits otherwise.

Malted.—A term applied to a preparation made

by combining the solids of cow's milk with the malt of grains. It is offered for sale in powdered or tablet form but is generally dissolved in water and consumed as a liquid. Frequently the tablets, however, are eaten directly.

Modified.—Milk whose constituents have been changed by proper blending or addition of other food substances to closely resemble in composition mother's milk. Modified milk is used chiefly for infant feeding.

Pasteurized.—Milk which has been subjected to sufficient heat to destroy most of the bacterial life contained without coagulating the casein or albumin of the milk. Pasteurizing temperatures commonly range from 140 to 180 degrees Fahrenheit.

Standardized.—A term indicating that milk has been blended with a richer milk in order to make it conform to a definite standard. For example, some Holstein milk is below the legal standard of certain cities in its butterfat content, but, if it is mixed with a small amount of cream or milk having more than the legal butterfat content, it can be legally sold. The mixing process is called standardization, and the milk mixed is referred to as standardized milk.

Starter Milk.—Milk used for the purpose of making a starter for butter making.

Sterilized.—Milk which has been scalded sufficiently to kill all bacterial life.

Tuberculin-tested.—A term applied to milk from a herd all of whose animals have successfully passed the tuberculin test. The milk itself has not been tested for the germs of tuberculosis and the term is, therefore, rather inaccurate, though commonly used.

Whole.—Milk from which no constituent has been taken or anything added; normal milk just as drawn from the cow.

Terms Used in Testing Milk and Its Products

Cow-testing.—The practice of determining by a series of accurate tests the profitable and the unprofitable cows in a dairy herd.

Cream Bottle.—A graduated bottle made of strong glass which is used for testing the amount of butterfat in cream.

Cubic Centimeter.—The unit of volume in the metric system which has been adopted as the standard for dairy testing. A linear centimeter is equivalent to 2.54 inches and the cubic centimeter is about one-sixteenth of a cubic inch.

Gram.—The unit of weight in the metric system. A gram is the weight of one cubic centimeter of distilled water at its maximum density.

Lactometer.—An instrument used for determining the specific gravity of milk. It consists of a bulb or float to the lower portion of which is attached a weight so that the lactometer will float upright in the milk. A scale on the neck of the instrument indicates the specific gravity directly.

Milk Sampler.—A small cup at the lower end of a long rod, used in taking an average sample from about the center of a large volume of milk.

Pipette.—A small glass tube with an enlargement near the center calculated to hold a definite amount of milk or other liquid. The lower end of the pipette is placed in the liquid and by applying suction at the upper end, the pipette may be filled to a definite mark just above the bulb.

Test.—A term applied to the percentage of butterfat in milk or its products. Adjectives such as light, dark and cloudy are used in connection with the word "test" to express the color of the butterfat column, the length of which determines the test.

Tester.—The centrifugal machine used in whirling the test bottles in making the test.

Terms Used in the Manufacture of Dairy Products

Carton.—The pasteboard package in which print butter is frequently put up for the retail trade.

Cheese Salt.—A very pure and easily soluble grade of salt used to salt cheese. The term "butter salt" is also sometimes used to define salt used in the manufacture of butter.

Color.—A harmless vegetable compound used to give butter and cheese a yellow color.

Curd.—The curdled or coagulated portion of milk which forms the basis of cheese.

Curd Mill.—A machine used for grinding curd for manufacture into cheese.

Curing Room.—A room kept at rather a low temperature in which cheese is held for several months in order that it may cure or ripen before going to the market.

Filler.—A substance such as gelatine, gum arabic or corn-starch used for the purpose of giving stiffness or body to ice-cream. Some fillers contain material calculated to give low grades of ice-cream the appearance of richness.

Overrun.—The difference between the amount of butter taken from the churn and the amount of butterfat from which the butter was made. As butter normally contains from ten to fifteen per cent. moisture, about three per cent. of salt and a considerable amount of casein and mineral matter, the amount of butter made exceeds the weight of the butterfat by from sixteen to twenty-two per cent. of the latter's weight. Overchurn is a term equivalent to overrun, but not in such frequent use.

Overworking.—The process injurious to the texture of butter which consists of working the butter too long a time.

Paraffining.—The application of a thin coating of paraffin to the inside of butter tubs or to the outside of cheese. Paraffining prevents the evaporation of moisture from dairy products.

Ripening.—The operation of souring milk or cream under the control of the butter maker; also a term used to describe the curing process in cheese.

Setting.—The first operation in preparing milk

for cheese making; for example, setting a vat of milk.

Weighing In.—The operation of receiving, weighing and keeping an account of the milk or cream received at a factory.

Yield.—The amount of cheese made from one hundred pounds of milk. The term "yield" is sometimes erroneously used for "overrun."

THE END

Index

ABUSE of pasteurization, 104
Accounts, dairy, 182
Advantages of dairy school assistance, 153
Age of cow important, 34
Animal diseases, information on, 155
Application of manure to fields, 178
Avoid cheap farm, 15
Ayrshires, characteristics of, 28–29

BABCOCK test, operation of the, 91
Bacteria in milk, 99
  tests for, 96
Balanced ration, the, 80
Barn-yard manure, its value, 177
Barrel churn, superiority of, 127
Basement barn objectionable, 42–43
Bookkeeping, its importance, 185
Bottled milk, the science of, 111
Box stalls, 50
Breeds, origin and characteristics, 22
  which are best? 20
Buildings, separate, 51
Bull pen, the, 51
Bulls, management of, 87
Business ability is essential, 13
Butter market not seriously threatened by oleo, 168
Butter markets, the three great, 159
Butter, terms applied to, 193
Buttermilk, 188

CAPS for bottles, different types of, 113
Care of the cream, 123
Careful investment, 15
Casein test, 93
Cattle feed is expensive fuel, 84
Ceiling, the cow barn, 46
Cement floor, 45
Certified milk and its production, 107
Characteristics of milk, 29
  of animals, 30
Cheapest and best foods, 84
Cheese factory, 17
  factories as markets, 161
  making, is it practical for the farmer? 133
Chemical fertilizers, 177
  preservatives, 105
Churning temperatures, 126
  the principle of, 128
Color of animal, 39
Common faults of cottage cheese, 135
Comparison of various grades of milk, 110

Concrete or wood structures, 44
Condensery, the danger of, 163
Considerations of minor importance, 39
Contents, 3, 4
Coöperation plan, organization of, 173
Cottage cheese, common faults of, 135
Cow barn, ceiling of, 46
Cream, care of, 123
Creamery, 17
Creamery promoter and his tactics, the, 169
Cross breeding, principle of, 32
Customary method of marketing butter, 174

DAIRY accounts, 182
  butter, how to make the best, 119
  farm fertility, 175
  markets, why they fluctuate, 157
  meetings, value of, 150
  organizations, 149
  publications, 154
  school assistance, advantages of, 153
  terms, 192
Dairying, 7
  for health and wealth, 188
Dairygrams, 190
Danger of the condensery, 163
Different types of caps, 113
Disposition of the animal, 38
Diversion for the hired men, 60
Doors and windows, 48
Dust problem solved, 101
Duties of State Dairy Department, 152

EDUCATIONAL and commercial value of dairy meetings, 150
Employer's obligations, 62
English cream cheese, 136
Errors not due to fault of test for tuberculosis, 145
Establishing a high class trade, 131

FACTORS which increase the value of dairy animals, 40
  which reduce the value of dairy animals, 41
Fat content of milk not affected by feed, 86
Feed for calves, 83
Feeding and managing the herd, 78
Fertility of dairy farm, how preserved, 175
  two theories of soil, 180
Fertilizers, 177
Few essential principles, a, 121
Forcing cows to make records, 35
Fraud of one-minute churns, 90
French cream cheese, 136

GETTING at the animal's net value, 184
Glass jar and the single service package, 112
Good milker, physical points of a, 37

Grass and hay, 16
Guernsey cow, the characteristics of, 24

HAND separators, 10
Hart casein test, 93
Health of animal, 31
Healthy calves from diseased cows, 147
Heat and cold as dairy agents, 100
Help problem, solution of the, 58
Helps for dairymen, 151
High-class trade, 131
Hindrances to successful milking, 67
Hired help, 18
Holsteins, 25
  characteristics of, 26
Housing and equipment, 42
How and why the dairy market fluctuates, 157
  to choose a dairy cow, 31
  to disinfect dairy premises, 146
  to make the best dairy butter, 119
  to select a dairy farm, 15

IMPORTANCE of both scientific and practical knowledge, 14
Information for dairymen, 151
Information on animal diseases, 155
Inspector's visits, 108
Is farm cheese making practical, 133

JERSEY cow, the, 22
Jersey cow, characteristics of, 23

LIGHT, 48
Locating a profitable side-line, 186

MAKING the starter, 125
Management of bulls, 87
Manure, application to fields, 178
Market milk prices, 158
Markets are constant, 9
  distance to, 16
Marketing butter, usual methods, 174
Milk, characteristics of, 29
  inspection and its purpose, 116
  records, an outline of, 183
  sanitation and care of, 98
  sediment test, 94
  strainer, the, 101
  terms applied to, 196
Milking machines—their practicability, 68
  —their proper management, 70
  —their mechanism, 71
  —difficulty of cleaning, 74
  —their saving of labor doubtful, 76
  cost of, 77
Milking, points on, 65
  rapidity of, 66
Minor considerations, 39
Moisture in butter, 93
Mortgaged farms, 16

NATIONAL DAIRY SHOW, 155

Natural feeds, 83
Net value of animals, 184
Nutritive ratio of feeds, 85
table of, 86

OBJECTIONS to pasteurized milk, 104
Obligations of employer, 62
of employee, 63
Oleo no serious menace to butter market, 168
Oleomargarine laws, the intent of, 167
why inferior to butter, 164
One-minute churn, 90
Opening era of better dairy butter, 120
Operation of the Babcock test, 91
Organs and tissues commonly affected by tuberculosis, 144
Organizations, dairy, 149
Organizing on the coöperative plan, 173
Outline of milk records, 183

PACKAGE as an index of its contents, the, 115
Palatable winter feeds, 82
Pasteurization and sterilization, 102
Pasture and corn, 16
Physical points of a good milker, 77
Points on milking, 65
Prices are inflated, 170
range of, 39
Principle of churning, the, 128
of cross-breeding, the, 32
Process of ripening the cream, 124
of working the butter, 130
Productiveness, 35
Proprietary preparations, 83
Proverbs, 190
Pure-bred stock always in demand, 33
Pure water, 79
Purpose of milk inspection, 116
Put back what you take away from the land, 176

QUARTERS for young calves, 50

RANGE of prices, 39
Rapid milking important, 66
Relative value of butter fat and oleo fat, 165
Rights of producers, inspectors, and consumers, 117
Ripening process, the, 124
Roughage, 81

SALTING and washing butter, 129
Sanitation and care of milk, 98
Science of bottled milk, 111
Seasonal rations, variety of, 81
Selection of a dairy farm, 15
Separate buildings for special operations, 51
Separation of milk by centrifugal force, 122
Separator, hand, 10
Services extended by dairy publications, 154
Silos and silage, 53
capacity of, 55
construction of, 56

Silos, foundation of, 57
Situation in the West, 160
Size of dairy farm, 18
Social life of the community, 19
Solving the dust problem, 101
the help problem, 58
Some common dairy terms, 192
Stalls and stanchions, 50
Sterilization, 102
Straining milk, 101
Stripping, 67
Study the animal's disposition, 38
Success in dairying, 12
Succulent feeds and pure water, 79
Summer and winter standards, 109
Sunlight as a germicide, 48
Superiority of the barrel churn, 127

TEMPERATURES for churning, 126
Terms applied to butter, 193
to cattle, 192
to milk, 196
used in the manufacture of dairy products, 201
used in testing milk and its products, 199
Test, the Hart casein, 93
milk sediment, 94
Tests for bacteria, 96
Testing milk for richness and purity, 89
Tests for solids in milk, 92
Three great butter markets, 159
Title, validity of, 19
Truth about the tuberculosis test, the, 138
Tuberculosis, 31
Two theories of soil fertility, 180

VALIDITY of title, 19
Value of heat and cold as dairy agents, 100
Value service available from Federal Government, 151
Varieties of cheese adapted for manufacture on farm, 134
Variety of seasonal rations, 81
Various grades of milk compared, 110
Ventilation, 49
Visits of inspectors are not announced, 108

WAGES, 59
Wagon roads, 16
Washing and salting the butter, 129
Water—quality and quantity, 16
Watering, 49
When cement floor is permissible, 45
When not to test for tuberculosis, 142
Why promoted creameries generally fail, 171
Windows and doors, size of, 48
Winter feeds, palatable, 82
Wood or concrete structures? 44
Work and wages, 59
Working butter, the process of, 130

YOUNG blood is best for foundation stock, 34
Young calves, quarters for, 50

ETIQUETTE

By Agnes H. Morton

There is no passport to good society like good manners. ¶ Even though one possess wealth and intelligence, his success in life may be marred by ignorance of social customs. ¶ A perusal of this book will prevent such blunders. It is a book for everybody, for the social leaders as well as for those less ambitious. ¶ The subject is presented in a bright and interesting manner, and represents the latest vogue.

LETTER WRITING

By Agnes H. Morton

Why do most persons dislike to write letters? Is it not because they cannot say the right thing in the right place? This admirable book not only shows by numerous examples just what kind of letters to write, but by directions and suggestions enables the reader to become an accomplished original letter writer. ¶ There are forms for all kinds of business and social letters, including invitations, acceptances, letters of sympathy, congratulations, and love letters.

QUOTATIONS

By Agnes H. Morton

A clever compilation of pithy quotations, selected from a great variety of sources, and alphabetically arranged according to the sentiment. ¶ In addition to all the popular quotations in current use, it contains many rare bits of prose and verse not generally found in similar collections. ¶ One important feature of the book is found in the characteristic lines from well known authors, in which the familiar sayings are credited to their original sources.

EPITAPHS

By Frederic W. Unger

Even death has its humorous side. ¶ There are said to be "sermons in stones," but when they are tombstones there is many a smile mixed with the moral. ¶ Usually churchyard humor is all the more delightful because it is unconscious, but there are times when it is intentional and none the less amusing. ¶ Of epitaphs, old and new, this book contains the best. It is full of quaint bits of obituary fancy, with a touch of the gruesome here and there for a relish.

PROVERBS

By John H. Bechtel

The genius, wit, and spirit of a nation are discovered in its proverbs, and the condensed wisdom of all ages and all nations is embodied in them. ¶ A good proverb that fits the case is often a convincing argument. ¶ This volume contains a representative collection of proverbs, old and new, and the indexes, topical and alphabetical, enable one to find readily just what he requires.

THINGS WORTH KNOWING

By John H. Bechtel

Can you name the coldest place in the United States or tell what year had 445 days? Do you know how soon the coal fields of the world are likely to be exhausted, or how the speed of a moving train may be told? What should you do first if you got a cinder in your eye, or your neighbor's baby swallowed a pin? This unique, up-to-date book answers thousands of just such interesting and useful questions.

A DICTIONARY OF MYTHOLOGY

By John H. Bechtel

Most of us dislike to look up a mythological subject because of the time required. ¶ This book remedies that difficulty because in it can be found at a glance just what is wanted. ¶ It is comprehensive, convenient, condensed, and the information is presented in such an interesting manner that when once read it will always be remembered. ¶ A distinctive feature of the book is the pronunciation of the proper names, something found in few other works.

SLIPS OF SPEECH

By John H. Bechtel

Who does not make them? The best of us do. ¶ Why not avoid them? Any one inspired with the spirit of self-improvement may readily do so. ¶ No necessity for studying rules of grammar or rhetoric when this book may be had. It teaches both without the study of either. ¶ It is a counsellor, a critic, a companion, and a guide, and is written in a most entertaining and chatty style.

HANDBOOK OF PRONUNCIATION

By John H. Bechtel

What is more disagreeable than a faulty pronunciation? No other defect so clearly shows a lack of culture. ¶ This book contains over 5,000 words on which most of us are apt to trip. ¶ They are here pronounced in the clearest and simplest manner, and according to the best authority ¶ It is more readily consulted than a dictionary, and is just as reliable.

PRACTICAL SYNONYMS

By John H. Bechtel

A new word is a new tool. ¶ This book will not only enlarge your vocabulary, but will show you how to express the exact shade of meaning you have in mind, and will cultivate a more precise habit of thought and speech. ¶ It will be found invaluable to busy journalists, merchants, lawyers, or clergymen, and as an aid to teachers no less than to the boys and girls under their care.

READY MADE SPEECHES

By George Hapgood, Esq.

Pretty much everybody in these latter days, is now and again called upon "to say a few words in public." ¶ Unfortunately, however, but few of us are gifted with the power of ready and graceful speech. ¶ This is a book of carefully planned model speeches to aid those who, without some slight help, must remain silent. ¶ There is a preliminary chapter of general advice to speakers.

AFTER-DINNER STORIES

By John Harrison

The dinner itself may be ever so good, and yet prove a failure if there is no mirth to enliven the company. ¶ Nothing adds so much zest to an occasion of this kind as a good story well told. ¶ Here are hundreds of the latest, best, brightest, and most catchy stories, all of them short and pithy, and so easy to remember that anyone can tell them successfully. ¶ There are also a number of selected toasts suitable to all occasions.

TOASTS

By William Pittenger

Most men dread being called upon to respond to a toast or to make an address. ¶ What would you not give for the ability to be rid of this embarrassment? No need to give much when you can learn the art from this little book. ¶ It will tell you how to do it; not only that, but by example it will show the way. ¶ It is valuable not alone to the novice, but to the experienced speaker, who will gather from it many suggestions.

THE DEBATER'S TREASURY

By William Pittenger

There is no greater ability than the power of skillful and forcible debate, and no accomplishment more readily acquired if the person is properly directed. ¶ In this little volume are directions for organizing and conducting debating societies and practical suggestions for all who desire to discuss questions in public. ¶ There is also a list of over 200 questions for debate, with arguments both affirmative and negative.

PUNCTUATION

By Paul Allardyce

Few persons can punctuate properly; to avoid mistakes many do not punctuate at all. ¶ A perusal of this book will remove all difficulties and make all points clear. ¶ The rules are plainly stated and freely illustrated, thus furnishing a most useful volume. ¶ The author is everywhere recognized as the leading authority upon the subject, and what he has to say is practical, concise, and comprehensive.

ORATORY

By Henry Ward Beecher

Few men ever enjoyed a wider experience or achieved a higher reputation in public speaking than Mr. Beecher. ¶ What he had to say on this subject was born of experience, and his own inimitable style was at once both statement and illustration of his theme. ¶ This volume is a unique and masterly treatise on the fundamental principles of true oratory.

CONVERSATION

By J. P. Mahaffy

Some people are accused of talking too much. But no one is ever taken to task for talking too well. ¶ Of all the accomplishments of modern society, that of being an agreeable conversationalist holds first place. Nothing is more delightful or valuable. ¶ To suggest what to say, just how and when to say it, is the general aim of this work, and it succeeds most admirably in its purpose.

READING AS A FINE ART

By Ernest Legouvé

The ability to read aloud well, whether at the fireside or on the public platform, is a fine art. ¶ The directions and suggestions contained in this work of standard authority will go far toward the attainment of this charming accomplishment. ¶ The work is especially recommended to teachers and others interested in the instruction of public school pupils.

SOCIALISM
By Charles H. Olin

Socialism is "in the air." ¶ References to the subject are constantly appearing in newspapers, magazines, and other publications. ¶ But few persons except the socialists themselves have more than a dim comprehension of what it really means. ¶ This book gives in a clear and interesting manner a complete idea of the economic doctrines taught by the best socialists.

JOURNALISM
By Charles H. Olin

What is news, how is it obtained, how handled, and how can one become a Journalist? ¶ These questions are all answered in this book, and detailed instructions are given for obtaining a position and writing up all kinds of "assignments." ¶ It shows what to avoid and what to cultivate, and contains chapters on book reviewing, dramatic criticism and proofreading.

VENTRILOQUISM
By Charles H. Olin

Although always a delightful form of entertainment, Ventriloquism is to most of us more or less of a mystery ¶ It need be so no longer. ¶ This book exposes the secrets of the art completely, and shows how almost anyone may learn to "throw the voice" both near and far. ¶ Directions for the construction of automatons are given as well as good dialogue for their successful operation. ¶ Fully illustrated.

CONUNDRUMS

By Dean Rivers

Conundrums sharpen our wits and lead us to think quickly. ¶ They are also a source of infinite amusement and pleasure, whiling away tedious hours and putting everyone in good humor. ¶ This book contains an excellent collection of over a thousand of the latest, brightest, and most up-to-date conundrums, to which are added many Biblical, poetical, and French conundrums.

MAGIC

By Ellis Stanyon

There is no more delightful form of entertainment than that afforded by the performances of a magician. ¶ Mysterious as these performances appear, they may be very readily learned if carefully explained. ¶ This book embraces full and detailed descriptions of all the well known tricks with coins, handkerchiefs, hats, flowers, and cards, together with a number of novelties not previously produced or explained. ¶ Fully illustrated.

HYPNOTISM

By Edward H. Eldridge, A. M.

There is no more popular or interesting form of entertainment than hypnotic exhibitions, and everyone would like to know how to hypnotize. ¶ By following the simple and concise instructions contained in this complete manual anyone may, with a little practice, readily learn how to exercise this unique and strange power.

WHIST

By Cavendish
Twenty-third Edition

"According to Cavendish" is now almost as familiar an expression as "according to Hoyle." ¶ No whist player, whether a novice or an expert, can afford to be without the aid and support of Cavendish. No household in which the game is played is complete without a copy of this book. ¶ This edition contains all of the matter found in the English publication and at one-fourth the cost.

PARLOR GAMES

By Helen E. Hollister

"What shall we do to amuse ourselves and our friends?" is a question frequently propounded on rainy days and long winter evenings. ¶ This volume most happily answers this question, as it contains a splendid collection of all kinds of games for amusement, entertainment, and instruction. ¶ The games are adapted to both old and young, and all classes will find them both profitable and interesting.

ASTRONOMY: The Sun and His Family

By Julia MacNair Wright

Can you tell what causes day and night, seasons and years, tides and eclipses? Why is the sky blue and Mars red? What are meteors and shooting stars? ¶ These and a thousand other questions are answered in a most fascinating way in this highly interesting volume. Few books contain as much valuable material so pleasantly packed in so small a space. ¶ Illustrated.

BOTANY:
The Story of Plant Life
By Julia MacNair Wright

The scientific study of Botany made as interesting as a fairy tale. ¶ It is better reading than such tales, because of the profit. ¶ Each chapter is devoted to the month of the year in which plants of that month are in evidence. Not only is the subject treated with accuracy, but there is given much practical information as to the care and treatment of plants and flowers. ¶ Illustrated.

FLOWERS:
How to Grow Them
By Eben E. Rexford

Every woman loves flowers, but few succeed in growing them. With the help so clearly given in this book no one need fail. ¶ It treats mainly of indoor flowers and plants—those for window gardening; all about their selection, care, soil, air, light, warmth, etc. ¶ The chapter on table decoration alone is worth the price of the book. ¶ While the subject of flowers is quite thoroughly covered, the style used is plain, simple, and free from all technicalities.

DANCING
By Marguerite Wilson

A complete instructor, beginning with the first positions and steps and leading up to the square and round dances. ¶ It contains a full list of calls for all of the square dances, and the appropriate music for each figure, the etiquette of the dances, and 100 figures for the german. ¶ It is unusually well illustrated by a large number of original drawings. ¶ Without doubt the best book on the subject.

ASTROLOGY

By M. M. Macgregor

If you wish to obtain a horoscope of your entire life, or if you would like to know in what business or profession you will best succeed, what friends you should make, whom you should marry, the kind of a person to choose for a business partner, or the time of the month in which to begin an enterprise, you will find these and hundreds of other vital questions solved in this book by the science of Astrology.

PHYSIOGNOMY

By Leila Lomax

How can we judge whether a man may be trusted to handle money for us? ¶ How can a woman analyze a man who would marry her? ¶ Partly by words, partly by voice, partly by reputation, but more than all by looks—the shape of the head, the set of the jaw, the line of the mouth, the glance of the eye. ¶ Physiognomy as explained in this book shows clearly how to read character with every point explained by illustrations and photographs.

GRAPHOLOGY: How to Read Character from Handwriting

By Clifford Howard

Do you know that every time you write five or six lines you furnish a complete record of your character? Anyone who understands Graphology can tell by simply examining your handwriting just what sort of a person you are. ¶ There is no method of character reading that is more interesting, more trustworthy, and more valuable than that of Graphology, and it is the aim of this volume to enable anyone to become a master of this most fascinating art.

CURIOUS FACTS

By Clifford Howard

Why do you raise your hat to a lady? and why are you always careful to offer the right hand and not the left? ¶ Is there a good reason for the buttons on the sleeve of your coat? ¶ How did your family name originate? ¶ Is it true that it takes nine tailors to make a man, and if so, why, forsooth? ¶ These and scores of equally interesting questions find answers here. Open it at any page and you will see something you have wanted to know all your life.

PRACTICAL PALMISTRY

By Henry Frith

The hand shows the man, but many who believe in palmistry have found no ready access to its principles. ¶ This little guide to it is complete, trustworthy, and yet simple in arrangement. ¶ With this book and a little practice anyone may read character surely, recall past events, and forecast the future. ¶ Fully illustrated.

CIVICS: What Every Citizen Should Know

By George Lewis

This book answers a multitude of questions of interest to everyone. ¶ It gives intelligent, concise, and complete information on such topics as the Monroe Doctrine, Behring Sea Controversy, Extradition Treaties, Basis of Taxation, and fully explains the meaning of Habeas Corpus, Free Coinage, Civil Service, Australian Ballot, and a great number of other equally interesting subjects.

LAW, AND HOW TO KEEP OUT OF IT

By Paschal H. Coggins, Esq.

Most legal difficulties arise from ignorance of the minor points of law. ¶ This book furnishes to the busy man and woman knowledge of just such points as are most likely to arise in every-day affairs, and thus protects them against mental worry and financial loss. ¶ Not only is this information liberally given, but every point is so explained and illustrated that the reader will not only understand the law on the subject, but cannot fail to remember it.

CLASSICAL DICTIONARY

By Edward S. Ellis, A. M.

All literature abounds in classical allusions, but many do not understand their meaning. ¶ The force of an argument or the beauty of an illustration is therefore often lost. ¶ To avoid this, everyone should have at hand a complete dictionary such as this. ¶ It contains all the classical allusions worth knowing, and they are so ready of access as to require little or no time in looking up.

PLUTARCH'S LIVES

By Edward S. Ellis, A. M.

Plutarch was the most famous biographer and one of the most delightful essayists who ever lived. ¶ To him we are indebted for an intimate acquaintance with many famous Greeks and Romans who made history and who still live. ¶ This book is a condensed form of the original "Lives." ¶ All the personages likely to be inquired about are mentioned, and what is told of them is just what one most wishes to know.